AF229427

OXFORD STATISTICAL SCIENCE SERIES

Symbolic Computation for Statistical Inference

D. F. Andrews

Department of Statistics, University of Toronto

and

J. E. Stafford

Department of Public Health Sciences and Department of Statistics, University of Toronto

OXFORD
UNIVERSITY PRESS

OXFORD
UNIVERSITY PRESS

Great Clarendon Street, Oxford OX2 6DP

Oxford University Press is a department of the University of Oxford.
If furthers the University's objective of excellence in research, scholarship,
and education by publishing worldwide in

Oxford   New York

Athens   Auckland   Bangkok   Bogota   Buenos Aires   Calcutta
Cape Town   Chennai   Dar es Salaam   Delhi   Florence   Hong Kong   Istanbul
Karachi   Kuala Lumpur   Madrid   Melbourne   Mexico City   Mumbai
Nairobi   Paris   São Paulo   Singapore   Taipei   Tokyo   Toronto   Warsaw

with associated companies in Berlin   Ibadan

Oxford is a registered trade mark of Oxford University Press
in the UK and in certain other countries

Published in the United States
by Oxford University Press Inc., New York

© D. F. Andrews and J. E. Stafford, 2000

The moral rights of the author have been asserted
Database right Oxford University Press (maker)

First published 2000

A catalogue record for this book is available from the British Library

Library of Congress Cataloging in Publication Data

Andrews, D.F. (David F.)
Symbolic computation for statistical inference/D.F. Andrews and J.E. Stafford.
p. cm. — (Oxford statistical science series; 21)
Includes bibliographical references and index. 1. Mathematical statistics–Data processing.
I. Stafford, J.E.H. (James E.H.) II. Title. III. Series.

QA276.4.A44 2000   519.5'0285—dc21   00-023769

ISBN 0 19 850705 4

Typeset by Newgen Imaging Systems (P) Ltd., Chennai, India
Printed in Great Britain
on acid-free paper by
Biddles Ltd, King's Lynn & Guildford

nihil dictum quad non dictum prius,
methodus sola artificem ostendit [1]

[1] As quoted in Merton [1965], *On the Shoulders of Giants*, from Burton [1651/2], *The Anatomy of Melancholy, What it is and All the Kinds, Causes, Symptoms, Prognostics and Several Cures of It in Three Partitions, with their several Sections, Members and Subsections, Philosophically, Medically, Historically Opened and Cut Up*, where it was usurped from Wecker è Ter. in Terence [161 BC], *Prologue* to *Eunuchus*.

Preface

This book was born of the conviction that symbolic computation could substantially extend the scope and depth of both the theory and methodology of statistical inference. It was raised in an exciting environment, challenged repeatedly with new problems and rewarded with the satisfaction of meeting these.

The development of statistical computing in the last third of this century has had a tremendous impact, freeing statisticians from the burden of calculation and opening up new fields of possible methods of analysis. There has been a substantial revolution in the role of statisticians.

The developments have been primarily in numerical computation and graphical display. Computational tools for analysis, design, and simulation have revolutionized the application of statistics bringing the methodology to a broad spectrum of workers in laboratories, industries, and business.

Symbolic computation offers as great a potential although, for reasons associated with the relative pace of the evolution of languages and machines, it comes now, some decades later.

The book summarizes roughly ten years of research into the use of symbolic computation applied to problems of statistical inference. The work began initially with the goal of developing tools for the machine computation of asymptotic expansions. Previously, the clerical aspects of this daunting task limited the advance of research.

We believed that much of the clerical work could be handled better by machine and were repeatedly challenged by friends who probably anticipated the result of their comments. Frequently we were met with phrases of the sort 'This is all very fine, but can you do it for arbitrary sample sizes?' or 'for arbitrary numbers of parameters?' or 'but can these procedures be used to prove the general result?' These challenging questions spurred us on and helped to guide us in our work.

As we learned more, the goals were extended and the interaction of theory and practice was played out. Inevitably the understanding of how things could be computed influenced our view of what things could be computed. The scope of the book is much broader than we initially envisioned. We hope the reader shares in at least part of the enjoyment we felt as we discovered these things.

We are grateful to those friends who, with their challenges and support, encouraged us in this work. This encouragement added to the sheer fun that comes from realizing how simple some previously complicated things can be.

This work owes much to the foundations and examples set by others. Indeed, the work began with the challenge to find the basic algorithms required to produce the results of McCullagh's book *Tensor Methods in Statistics* (Chapman and Hall 1987). Subsequently we studied *The Bootstrap and Edgeworth Expansion* by Hall (Springer-Verlag 1982) and *An Introduction to the Bootstrap* by Efron and Tibshirani (Chapman and Hall 1993). Our goal was to free the statistician from the mundane clerical aspects of research in these areas. The tools that resulted from this work provide, we believe, the potential for much that is new and exciting.

As noted above we were aided and encouraged by many friends including Professors W. S. Kendall, N. Reid, and S. Stigler. Anthony Davison contributed greatly to the final version of the monograph with his meticulous proof reading of most chapters and uncompromising comments on content. David Bellhouse, Duncan Murdoch, and Robert Philips provided similar assistance for various chapters. We have benefitted from many discussions with D. Binder, M. Hidiroglou, K. Knight, P. McCullagh, P. Mykland, D. Wallace, and H. Wynn among many others. R. Tibshirani was particularly influential providing continuous support in a variety of ways.

A large number of students have collaborated on related developments. These include Sookhee Yun, George Sampson, Brent Knowles, Steve Drekic, Alison Weir and most notably, Yong Wang who courageously joined us in the early work related to likelihood. We wish to thank all students at The Universities of Western Ontario, Toronto, Chicago, Stanford, and The Swiss Federal Institute of Technology Lausanne who attended courses during the development of the book. Their comments have been most helpful.

In addition, we thank M. Nandi, E. Johnston, R. Lawrence, and all those at OUP who so carefully contributed to the publication of this book.

We are grateful to family members for their patience, support, and occasionally their meaningful contributions to content. These include Linda, Duncan, and Stuart Andrews. Thanks to Maria-Luisa Gardner for providing expert advice on the 'rules of punctuation'. Gabriela Stafford said the book was funny while her younger sister Lucia thought the pages tasted good.

We are grateful for the support of this research by the Natural Sciences and Engineering Research Council of Canada and by Statistics Canada.

<table>
<tr><td>Toronto</td><td style="text-align:right">D.F.A.</td></tr>
<tr><td>November 1999</td><td style="text-align:right">J.E.S.</td></tr>
</table>

Contents

1
Introduction

The theoretical basis of statistical methods is mathematics. The power of mathematics comes from the generality of its results. Because the relations are so general, only a few are required for a great variety of calculations. The power of computer algebra lies in rendering these general relations in a form for use by machine. As a result, a very limited amount of programming makes possible algebraic computation for a great variety of problems. Symbolic algorithms are to mathematics as numerical algorithms are to arithmetic.

This is a book about the use of such algorithms in statistics. By symbolic computation, or computer algebra, we refer to the calculation of *algebraic expressions* such as $E[X^2]/n - (E[X]/n)^2$ for the variance of an average, rather than the numerical evaluation of this expression for a particular distribution.

The development of statistical languages based on advances in graphical and numerical algorithms in the last three decades has revolutionized the methodology and practice of statistics. The development of symbolic statistical algorithms began much later, but we anticipate they will have a comparable impact on both theoretical and applied statistical research.

1.1   What do we mean by symbolic computation?

By symbolic computation we refer to the use of computer languages and procedures to manipulate expressions replacing the statistician's use of a pencil and paper for this purpose. As with numerical computation, the use of machines greatly extends the capability of the statistician and makes new methodology possible. Moreover, the results are achieved faster, are reproducible, and, with proper algorithms, correct.

The use of computers to perform algebraic calculations is the next frontier. Calculators have diminished the importance of teaching the detailed mechanics of methods for long division in schools. Numerical algorithms have replaced the distracting arithmetic of least-squares and analysis of variance in courses in statistics. Monte Carlo procedures have made possible the evaluation of high-dimensional integrals and have led to the development of new statistical methods including bootstrap procedures based on resampling. Symbolic computation has similar potential: to handle complex algebraic calculations of probabilities, expectations, densities, asymptotic expansions, cumulants, and their estimates to make advances in statistical methodology.

What calculations can be automated? How can we automate them? What new concepts are required? These questions are addressed in the context of problems encountered in statistical research and chiefly, though not solely, in asymptotic theory where the computational burden can be limiting. We focus on the use of computer algebra as a research tool and forego discussion of implementation in a specific language since it requires the user to be familiar with that language.

The objective of this book is to pass to a machine the mechanical processes of algebraic calculation, thus allowing the user to concentrate on the conceptual aspects of a problem. Algorithms are rendered in a form that can be implemented in any suitably sophisticated language. We believe experience is enhanced by the method of trial and we encourage the reader to implement these algorithms (there are not too many) or extend those in the associated Mathematica notebooks. They contain procedures for the computation of all the expressions in this book (and a great many others). Our interest has been primarily in the definition and structure of the operators. Better implementations can no doubt be found.

With the great increase in computational power in the last three decades have come parallel developments assisting the user in doing mathematics by computer to the point where there are now numerous systems available including Mathematica, Maple, Macsyma, and Reduce. All of these environments include a core of common algorithms for basic computations. The typical operations automated include: addition, multiplication, integration, differentiation, plotting, etc. However, the greater potential of these environments lies in the ability to effectively teach the computer to perform an algebraic calculation for which an algorithm exists.

The tedium and risk of error in a human calculation arises when the same rule is used not just once or twice, but an overwhelming number of times. For example, differentiating a moment generating function to produce moments involves the repeated use of the product rule for differentiation. Symbolic computation allows the user to replace the iterative algebra with one rule. Betraying our preference for Mathematica, we define procedures by sets of rules. Doing so requires thinking deeply about a calculation to identify its fundamental structure. Typically, one rule replaces many operations, making a computer algebra operator a tremendous labour-saving device.

Consider the problem of computing moments for a multinomial distribution with n trials and k categories, $\{X_1, \ldots, X_k\} \sim Mult(n, k)$. Operators reflect the simple dependence of this calculation on a set of indices. For the set $s \subset S = \{1, \ldots, k\}$ we let $\mathbf{X}_s$ denote $\prod_{i \in s} X_i$ and t_s, ∂_s are defined to evaluate as $\{\ldots, t_i, \ldots\}$ and $\prod_{i \in s} \partial/\partial t_i$, respectively. Two rules suffice:

$$E[\mathbf{X}_s] = [\partial_s M(t_S)]_{t \to 0},$$

$$M(t_S) = \left[\sum_{i=1}^{k} p_i \exp(t_i) \right]^n. \tag{1.1}$$

The first defines expectation through differentiation for any distribution, while the second gives the moment generating function for this example. The function

$\partial/\partial t_i$ computes derivatives and, while such functions are easily specified, they are typically implemented in a language.

Using these definitions, joint moments may now be computed as

$$E[X_1 X_2 X_3] \longrightarrow (-2 + n)(-1 + n)n p_1 p_2 p_3,$$

where the result has been simplified using $\sum p_i = 1$. Throughout this book, the notation '$\longrightarrow$' denotes symbolic computation where the input on the left-hand side produces the computed result on the right-hand side.

1.2 Expanding the toolbox

The example above is typical of what may be calculated directly in a language for symbolic computation. However, we can extend this simple moment calculation to a problem that is central to many of the algorithms described in this book.

Suppose our interest lies not in the evaluation of a moment $E[X_1 X_2 X_3]$, but rather in the expression of the moment in terms of cumulants, the coefficients in the expansion of $K(\cdot) = \log\{M(\cdot)\}$. Similar to moments, the cumulants can be obtained by evaluating the derivatives of the cumulant generating function $K(\cdot)$ at 0

$$[\partial_s K(t_S)]_{t \to 0}$$

where s, S, and t have the same definitions as before. The first few of these are the mean, variance, skewness, etc. Now using the definition of $K(\cdot)$, we may define the moment generating function as

$$M(t_S) = e^{K(t_S)}. \tag{1.2}$$

With this definition, the procedure for computing moments, (1.1), gives a generic result:

$$E[X_1 X_2 X_3] \longrightarrow \kappa_1 \kappa_2 \kappa_3 + \kappa_3 \kappa_{1,2} + \kappa_2 \kappa_{1,3} + \kappa_1 \kappa_{2,3} + \kappa_{1,2,3},$$

where the derivatives of $K(\cdot)$, when evaluated at 0, are set to display as κ_1 for the mean $E[X_1]$, $\kappa_{1,2}$ for the covariance of X_1, X_2, and so on.

The important feature is the structure that may be described as a sum over $\mathcal{P}$, all partitions of the index set $\{1, 2, 3\}$, where $\mathcal{P} = \{(1|2|3), (1|23), (13|2), (12|3), (123)\}$. This structure arises repeatedly in very different calculations and it is this common structure that our algorithms exploit. That is, simple algorithms have broad applications.

The general partition structure is clear from the multivariate calculation, but not from the univariate simplification where $X_3 = X_2 = X_1$ and

$$E[X_1^3] = \kappa_1^3 + 3\kappa_1 \kappa_{1,1} + \kappa_{1,1,1}$$

or from the multinomial calculation. However, the same algorithm is used in each case. Often, the simple common structure is transparent in the multivariate case but obscured in the *simpler* univariate case. This is a general theme that is underscored in McCullagh (1987), though, as he admits, not widely acknowledged. Our efforts support this theme. All algorithms are defined for the multivariate case because of the simple structure present. Simplification to the univariate case is designed to speed up computations, but does not affect the basic code.

1.3 What's in this book?

This book concerns symbolic algorithms and their application in statistical inference. The chapters that follow can be roughly grouped into these two categories, with Chapters 2–5 and 10 being devoted to the development of algorithms and the remaining to their application. The development of algorithms begins in Chapter 2 with the implementation for some standard problems from a first course in mathematical statistics. By the end of this chapter, it is evident that coping with more challenging problems will require careful development of algorithms.

Chapter 3 reports the most ambitious material in the text and acts as a litmus test for the reader in terms of what s/he will gain. Computer algebra environments, and especially the one we use, rely heavily on the use of lists. So, the challenge is not just to implement complex calculations but to think of these calculations in a radically different way. Namely, we need to think about mathematical expressions as lists. While this may appear to be an unnecessary *complication*, it is in fact a beautiful *simplification*. For example, many calculations involve partitions of lists, so a partitioning algorithm is flexible if expressions are represented as lists. Chapters 4 and 5 present algorithms for asymptotic expansions and coupling these with the algorithms of Chapter 3.

The remaining chapters report applications of algorithms to different areas of statistical methodology. These include likelihood inference, the bootstrap, survey sampling, Edgeworth and saddle-point calculations. Each application is accompanied by what we think is a sufficient discussion of the foundation of the methodology without delving into overwhelming detail. The automation of an algebraic calculation precludes the detail of how it was done in the past by hand. However, the background details are crucial to appreciating the automation. Some further development of algorithms occurs when needed. The major contribution being simplification procedures for generic multivariate calculations.

The single most important theme throughout this book is the need for the reader to think broadly about the common structure of calculations. The book is intended to complement the current literature. It does not present the basis for the theory, it represents the theory in a form for use by machine. For example, we provide and discuss powerful algorithms for cumulant calculations such as those in McCullagh (1987), but the background detail justifying the use of cumulants is left to that source.

1.4 A problem

In this problem, we ask the reader to utilize a computational trick that allows us to write cumulants in terms of moments without changing the rule (1.1). This exercise has no purpose other than to emphasize one recurrent theme of the book: by thinking about the structure of a calculation, we can identify what is has in common with other calculations and thereby create general algorithms.

Both $M(t)$ and $K(t)$ have Taylor expansions:

$$M(t) = 1 + \mu_1 \cdot t + \frac{1}{2}\mu_2 \cdot t \cdot t + \frac{1}{3!}\mu_3 \cdot t \cdot t \cdot t + \cdots,$$

$$K(t) = \kappa_1 \cdot t + \frac{1}{2}\kappa_2 \cdot t \cdot t + \frac{1}{3!}\kappa_3 \cdot t \cdot t \cdot t + \cdots,$$

where the coefficients in each case are arrays with dimension denoted by the subscript. The expansions have the same structure, with the exception that the leading term of $M(t)$ is 1, which, for the purpose of computing moments and cumulants, is irrelevant. Thus, treating the expansions as equivalent, or rather M/K and μ/κ as interchangeable, define $M(t) = \log\{1 + K(t)\}$ and use (1.1) to write the cumulants in terms of moments to compute κ_{X_1,X_2,X_3} in terms of the moments.

Implementations of this and subsequent chapters may be found at

http://www.utstat.utoronto.ca/david/SCSI.html

2
Probability and random variables

The basic definitions and relations used in statistics may be specified in a language for symbolic computation and then used for calculation. Programming a machine to do symbolic computation in statistics is similar to teaching undergraduates. Although it is not as much fun, the machine has a superhuman attention span. The following sections present algorithms for many of the calculations faced by undergraduates. It is surprising how little needs to be specified for these calculations. As with undergraduates, we assume the language of implementation has (only) the basic prerequisite operational skills in algebra, arithmetic, and calculus, comparable to those required to complete a first-year course in calculus. We assume therefore that procedures for matrix multiplication, inversion, and simple integration are available. Aside from these, all of the definitions and procedures required are indicated in the text. In almost all cases, these are set out in displays.

New notation for familiar operators reminds the reader that these are not merely abstract notions but are intended for explicit implementation as algorithms.

The system used for the preparation of this book, Mathematica, applies definitions to expressions to create a new expression. In other words it transforms expressions. Thus, for example, letting $\cup$ denote a union operator for sets, the rule given below states that the union of A with A is just A. It is specified by:

$$\cup[A, A, \ldots] = \cup[A, \ldots],$$

where A is an arbitrary event. The precise specification of the rule for Mathematica involves some symbols peculiar to that system. Definitions are presented here in a more generic fashion.

This rule causes the expression $\cup[B, C, B]$ to be simplified to $\cup[B, C]$. This computation is displayed as:

$$\cup[B, C, B] \to \cup[B, C].$$

As in an undergraduate course, this development is cumulative; definitions implemented in one section will be applied to expressions in subsequent sections.

The following sections of this chapter illustrate how fundamental statistical concepts may be implemented. Distinctive data structures for basic components are used to simplify both programming and computation.

2.1 Probability

Like many undergraduate courses in theoretical statistics, this chapter begins with probability. The basic concepts of the subject are described and definitions relating these are implemented. Only a small number of definitions are required for the computation of a rather large class of problems, as indicated by the examples.

2.1.1 OUTCOMES, SAMPLE SPACE, AND EVENTS

A typical introduction to probability begins with the definition of outcomes, a sample space, and events. These are the objects used to define probability. They are like assumptions, to be accepted; no implementation is involved here.

An outcome is the result of a trial or experiment. The sample space is the set of all possible outcomes. It is commonly denoted by Ω. The empty set is denoted by $\emptyset$.

Events are sets of outcomes. In the following text, these will be denoted by $A, B, \ldots$, although this restriction is not part of the implementation. The complement of the event A is represented by $\overline{A}$. A precise definition of probability restricts events to be elements of a σ-field. This is an important theoretical restriction. It is implemented here by assumption: it is assumed that all events belong to a σ-field. This means that all complements, unions, and intersections formed from the specified events are themselves events, members of the σ-field. (Although important, this theoretical consideration has never troubled statisticians, nor have they suffered for focusing on other matters.)

2.1.2 UNIONS AND INTERSECTIONS OF EVENTS

The set operations of union, $\cup$, and intersection, $\cap$, are used to construct events from more basic components. Although these operations are commonly implemented in languages for symbolic computation, they are implemented here as an exercise to illustrate the simplicity of the definition of such general procedures.

The union of events $A, B, \ldots$ is represented by $\cup[A, B, \ldots]$ and their intersection by $\cap[A, B, \ldots]$. The rules defining the operation of $\cup$ an $\cap$ are implemented by:

$$\cup[A] = A, \quad \cap[A] = A,$$

$$\cup[\cup[A, \ldots_1], \ldots_2] = \cup[A, \ldots_1, \ldots_2], \quad \cap[\cap[A, \ldots_1], \ldots_2] = \cap[A, \ldots_1, \ldots_2],$$

$$\cup[A, A, \ldots] = \cup[A, \ldots], \quad \cap[A, A, \ldots] = \cap[A, \ldots],$$

$$\cup[A, \overline{A}, \ldots] = \Omega, \quad \cap[A, \overline{A}, \ldots] = \emptyset,$$

$$\cup[\Omega, \ldots] = \Omega, \quad \cap[\Omega, \ldots] = \cap[\cdots],$$

$$\cup[\emptyset, \ldots] = \cup[\cdots], \quad \cap[\emptyset, \ldots] = \emptyset,$$

where $\ldots$ and $\ldots_i$ denote 0, 1, or more arguments. The subscripts $\ldots_i$ are used to identify identical arguments. The order of the arguments does not matter.

One rule defines the distribution of $\cap$ over $\cup$:

$$\cap[\cup[A,\ldots_1],\ldots_2] = \cup[\cap[A,\ldots_2],\cap[\cup[\ldots_1],\ldots_2]].$$

The rule is applied recursively whenever $\cup[\ldots_1]$ or $\ldots_2$ is a union of more than one event. A similar rule defines the second distributive law.

The commutative nature of the operations is seemingly not required here and so is not implemented. The operation of union and intersection will be applied automatically to any subsequent expressions.

2.1.3 PROBABILITY

The probability of an event A is denoted by $P[A]$. The conditional probability of A given B is denoted by $P[A, B]$. At this point there are only two special events, namely Ω and $\emptyset$, and only two functions of events, namely $\cup$ and $\cap$. Probability is defined for these by four axioms or rules, implemented as follows:

$$P[\Omega] = 1,$$

$$P[\emptyset] = 0,$$

$$P[\cup[A,\ldots_1],\ldots_2] = P[A,\ldots_2] + P[\cup[\ldots_1],\ldots_2] - P[\cap[A,\cup[\ldots_1]],\ldots_2],$$

$$P[\cap[A, B]] = P[A, B]P[B].$$

$$(2.1)$$

In the third rule, $\ldots_2$ is an optional argument. Thus, the probability of a union is defined for both the conditional and unconditional events.

The rules for probability, specified above, may be used to compute probabilities of unions. For example, the union of three events, given by the left-hand side below, results in the expression on the right:

$$P[\cup[A, B, C]] \rightarrow P[A] + P[B] + P[C] - P[\cap[A, B]]$$
$$- P[\cap[A, C]] - P[\cap[B, C] + P[\cap[A, B, C]].$$

This result is usually demonstrated by proof in an undergraduate course. Here it results from a single rule that is iterated by the computer. The lesson here is that most proofs just involve iterating rules until the final expression appears. Machines, of course, have the advantage of being able to iterate rules very quickly and without error. In Chapter 3, we expand on the theme of computer algebra proofs as a means of deriving correct algorithms.

2.1.4 PARTITIONS

In some situations, an event may be split into a number of disjoint, or non-intersecting, events. The decomposition is called a partition. For example, the sample space may be split into two disjoint events A and $\overline{A}$. When an event is a component of a partition, the partition may be used in computing probabilities, as shown in the example below.

We define a partition as the union of disjoint events and denote it by $\mathcal{P} \equiv \mathcal{P}[\mathcal{P}_1, \mathcal{P}_2, \ldots]$. Here the operator $\cup$ has been replaced by $\mathcal{P}$ to indicate that the elements of this union have a special property. The jth element of the partition is denoted by $\mathcal{P}_j$. Because events of a partition, by definition, do not intersect, the rules for $\cap$ may be extended:

$$\cap\,[A, \mathcal{P}] = \cup[\cap[A, \mathcal{P}_1], \cap[A, \mathcal{P}_2], \ldots],$$

$$\cap\,[\mathcal{P}_j, \mathcal{P}_k, \ldots] = \emptyset \quad \text{if } j \neq k.$$

These rules result in the law of total probability being implicitly invoked (without proof) any time P is applied to the intersection of an event with a partition. For example, if $\mathcal{P}$ is defined such that $\mathcal{P}_1 = A$, $\mathcal{P}_2 = B$, and $\mathcal{P}_3 = C$, then

$$P[\cap[E, \mathcal{P}]] \longrightarrow P[A]P[E, A] + P[B]P[E, B] + P[C]P[E, C].$$

To further exploit the structure of partitions, the rules for probability are augmented by writing the last rule (2.1) for partitions as

$$P[\mathcal{P}_j, A] = P[\cap[\mathcal{P}_j, A]] / P[\cap[A, \mathcal{P}]].$$

These few rules are sufficient for the computation of conditional probabilities including those often associated with Bayes' theorem. Their application is illustrated with the following example.

2.1.5 EXAMPLE: *TB* OR NOT *TB*

Imagine that individuals in a population could be classified as carrying or not carrying a disease, tuberculosis for example. This is a convenient, if incorrect, simplification of life. Consider an *experiment* in which a person is drawn at random from this population. The implied sample space, Ω, is the population. Let TB denote the event that a person drawn from the population has the disease, and let $\overline{TB}$ denote the event that they do not have the disease. Thus, if $\mathcal{P}$ is defined such that $\mathcal{P}_1 = TB$ and $\mathcal{P}_2 = \overline{TB}$, then $\mathcal{P}$ is a partition of the population Ω.

Experience may give some indication of the probability that a person, drawn at random from the population, does or does not carry the disease. Let $P[TB]$ and $P[\overline{TB}]$ denote these two probabilities. Similarly, experience with a diagnostic test for the disease, may give some indication of the probabilities of a positive test for people with and without the disease. Let $P[pos, TB]$ and $P[pos, \overline{TB}]$ denote these probabilities.

It is of interest to compute $P[TB, pos]$, the probability that a person, drawn at random from the population, carries the disease, given that they have tested positive. This is computed from the above definitions which cause the probability $P[\mathcal{P}_1, pos]$ to produce the right-hand side below:

$$P[\mathcal{P}_1, pos] \to \frac{P[TB]P[pos, TB]}{P[TB]P[pos, TB] + P[\overline{TB}]P[pos, \overline{TB}]}.$$

Once again, a well-known result, namely Bayes' rule, is implicit without any special knowledge or implementation. This and the implicit nature of the law of total probability are again examples of 'proofs' that simply involve the iteration of one or more rules. Once such rules are defined for operators, they are automatically iterated in calculations and it seems that the special results become 'redundant'. However, one must be cautious in drawing such conclusions. While it is true that the results are no longer needed for correctness, they often improve efficiency. That certainly would be true here if the calculation was large enough. Nevertheless, this apparent redundancy of cherished results in basic probability should deepen our understanding of the results and proofs in general. This would not be possible without the use of computer algebra—well, not for most people anyway. A more extensive use of partitions is illustrated by the genetics example below.

2.1.6 INDEPENDENT EVENTS

In many situations certain events are assumed to be statistically independent. Independent events simplify calculations involving intersections and so, as with elements of a partition, we reserve a special notation for the intersection of such events. The intersection of independent events $A, B, \ldots$ is denoted by $\mathcal{I}[A, B, \ldots]$. For such intersections, we have the following rule for P:

$$P[\mathcal{I}[A, B, \ldots]] = P[A]P[B]\cdots$$

A collection of partitions, $\mathcal{P}^1, \mathcal{P}^2, \ldots$ is said to be independent if all collections, $\mathcal{P}^1_i, \mathcal{P}^2_j, \ldots$ made up of one event from each partition, form an independent set of events. The set of all such collections is itself a partition. This set is generated by a single rule for the intersection of independent partitions:

$$\mathcal{I}[\mathcal{P}^1, \mathcal{P}^2, \ldots] = \mathcal{P}[\ldots, \mathcal{I}[\mathcal{P}^1_i, \mathcal{P}^2_j, \ldots], \ldots].$$

For example, for two independent events A and B we can generate a third partition from the two independent partitions $\mathcal{P}^1 \equiv \mathcal{P}[A, \overline{A}]$ and $\mathcal{P}^2 \equiv \mathcal{P}[B, \overline{B}]$,

$$\mathcal{I}[\mathcal{P}^1, \mathcal{P}^2] \to \mathcal{P}[\mathcal{I}[A, B], \mathcal{I}[A, \overline{B}], \mathcal{I}[\overline{A}, B], \mathcal{I}[\overline{A}, \overline{B}]].$$

2.1.7 A GENETIC EXAMPLE

Under assumptions of random mating and those of Mendelian genetics, the distribution of genetic types in a population stabilizes with the first generation. The *laws* on which this is based are:

- genetic characteristics of parents are independent under random mating
- independently, each parent passes one of two alleles to a child, each with probability $\frac{1}{2}$.

To derive the result, the *laws* of Mendelian genetics and of random mating are specified in terms of partitions and independence. The definitions result in computed probabilities for the genetic types for each generation.

Alleles, genes, and genotypes   Genes may take on different forms called alleles: $a_1, a_2, \ldots$ Chromosomes contain two alleles at each site or locus: one came from the mother and one from the father, although the source cannot be determined easily by observation. Thus, a genotype is an unordered pair: $g[a_i, a_j]$.

Partitions of genotypes   The problem is to compute the probability that a child of a random mating from some population has a particular genotype. This probability depends on the probabilities of the various possible genotypes of the parents. These genotypes play a role similar to the disease status $TB, \overline{TB}$ of the previous example. Here, as there, it is useful to represent the collection of all genotypes as a partition.

The partitions corresponding to genotypes of individuals do not need to be defined. Genotypes of mothers, fathers, and children represent different events and therefore should have different labels. However, the labels are not necessary for computation, and efficiency is improved by avoiding as much notation as possible.

For simplicity of exposition, the discussion below is for a locus with only two alleles: a, A. The possible genotypes of the mother, father, and the child are each given by the same partition

$$\mathcal{P} \equiv \mathcal{P}[g[a, a], g[a, A], g[A, A]].$$

Under the assumption of random mating, the partition giving all the possible genotype combinations for the parents is given by $\mathcal{I}[\mathcal{P}, \mathcal{P}]$ whose nine elements of the form $\mathcal{I}[g_1, g_2]$, explicitly indicate the independence of the genotypes for the mother and father. Such an intersection produces a child and, for this wonderful occurrence, we are interested in the probability of each genotype for the child.

Efficiency in computation   This may be accomplished by the command $P[\cap[g, \mathcal{I}[\mathcal{P}, \mathcal{P}]]]$ which invokes the rules for $\mathcal{I}, \mathcal{P}, \cap, P$ and ultimately derives the law of total probability... if one waits long enough. The difficulty is that the calculation implicitly derives the inclusion–exclusion rule for the union of nine events. The resulting expression has 511 terms but 502 of them evaluate to 0. While it fascinates us that computers can be taught to derive results like the law of total probability, doing so in every calculation is not in the best interest of efficiency. Thus, such rules should be explicitly defined to improve the efficiency of larger calculations.

Population probabilities   The distribution of genotypes in the original population is specified by

$$P[g[a, a]] = p, \quad P[g[a, A]] = 2q, \quad P[g[A, A]] = r.$$

(Note that this simplification is not reasonable in cases where the survival of offspring is related to their genotype as, for example, used to be the case for Duchenne muscular dystrophy.) Since the probabilities of the genotypes add to 1, the expression is further simplified by setting

$$r = 1 - p - 2q.$$

Independently, each parent passes one of two alleles to a child with equal probability. Letting $g_c = g[c_1, c_2]$, $g_f = g[f_1, f_2]$, and $g_m = g[m_1, m_2]$ be the genotypes of the child, father, and the mother, respectively, the Mendelian rule for computing the conditional probability of g_c, given g_f and g_m, is the coefficient of g_c in the distribution $(1/4)(g[f_1, m_1] + g[f_1, m_2] + g[f_2, m_1] + g[f_2, m_2])$. These give the conditional probabilities $P[g, \mathcal{I}[g_f, g_m]]$.

The unconditional probabilities for the child genotype may now be computed as

$$P[\cap[g[a, a], \mathcal{I}[\mathcal{P}, \mathcal{P}]]] \to (p + q)^2,$$

$$P[\cap[g[a, A], \mathcal{I}[\mathcal{P}, \mathcal{P}]]] \to 2(1 - p - q)(p + q),$$

$$P[\cap[g[A, A], \mathcal{I}[\mathcal{P}, \mathcal{P}]]] \to (1 - p - q)^2.$$

These represent the probabilities of genotypes of the next generation.

2.1.8 HARDY–WEINBERG EQUILIBRIUM

If *these* values are substituted for p, $2q$, and r, the resulting probabilities are the same!. This is known as the Hardy–Weinberg equilibrium. The probabilities of genotypes have stabilized after the first generation.

2.2 Random variables and their properties

A typical introductory course in mathematical statistics begins with the definition of a probability space and proceeds to concepts of random variables, distributions, generating functions, an expected value operator, etc. We can construct a 'package' to automate the routine calculations typically done by hand in a first course. The two fundamental constructs required are random variables and distributions or laws. All other objects and operators follow from these. Objects like densities, likelihoods, and generating functions have mathematical forms that are unique for a law but require a random variable for evaluation. Symbolic programming allows for these subtle abstract distinctions in an explicit way.

In what follows, we only define two operators: one for identifying random quantities and one for computing expectation. Objects like laws, densities, and generating functions are assumed to be user-defined. Therefore, the 'package' requires only seven essential lines of code. However, we can use this package to discuss, and/or implement, Bartlett's identities, model misspecifications, the delta method, symbolic Newton–Raphson, and Bartlett's correction for likelihood ratio statistics. These fairly advanced topics require only simple tools.

2.2.1 RANDOM VARIABLES

Statistics deals with numbers. The numbers are observed values of random variables and often no distinction is made between the two. So random variables are central to statistics and in a large sense characterize statistical methodology. A random variable is defined as a real-valued function on a sample space. This definition implies that any function of a random variable is itself a random variable. It follows that, if random variables are represented by a distinctive form, then these variables and their functions may be readily identified. It will be possible to develop an *operational* rule to determine if an expression involves and, therefore, is a random variable. This determination is important because many statistical quantities have properties which depend on whether or not random variables are involved. For example, if c is a constant, not a random variable, and X a random variable, then expectation satisfies $E[cX] = cE[X]$.

In print, random variables are often identified by the case of the symbol used, as in the above cX. However, this distinction is immediately lost when one considers the *constant*, $E[X]$. Here we identify random variables by the operator $\mathcal{R}$ defined to return

$$\mathcal{R}(\cdot) = \text{False},$$

except for overiding definitions. For example, X is defined to be a random variable by writing

$$\mathcal{R}(X) = \text{True},$$

otherwise X is not a random variable. In addition, the rule

$$\mathcal{R}(g[\cdot]) = \mathcal{R}(\cdot)$$

ensures functions of a random variable are identified as random variables. Exceptions to this rule, such as $E[X]$, must be made explicit, for example, by writing

$$\mathcal{R}(E[\cdot]) = \text{False}.$$

2.2.2 EXPECTATION

With random variables identified, it is possible to set down some rules for an expectation operator, E. It is initially defined as a linear operator which preserves constants:

$$E[X + Y] = E[X] + E[Y],$$

$$E[aX] = aE[X],$$

$$E[a] = a.$$

The last two rules are invoked when a is not a random variable, a condition that can be checked using $\mathcal{R}$. Further rules will be needed for the evaluation of $E[X]$ for a particular distribution. In addition, cases where X is a product of sums of unspecified length n motivate the core material of this book. While we are familiar with manipulating sums of this type, computers are not, and it is not immediately obvious how to teach them. However, in doing so, we lay bare a structure that is common to many seemingly unrelated calculations. Given this, the objective of later chapters, in particular Chapters 3 and 4, is how to design operators to efficiently exploit this common structure.

2.2.3 LAWS AND PARAMETERS

The stochastic behaviour of a random variable X is described by its distribution which may belong to a larger family of distributions we identify by a law. The Gaussian, binomial, and Poisson are examples of laws. The function $\mathcal{L}$ is used to identify the law of a random variable. For example, the distribution of Y is defined to be Poisson by

$$\mathcal{L}(Y) = \text{Poisson}.$$

$\mathcal{L}$ has no defining rules other than ones like this.

In parametric inference, families of distributions are indexed by a parameter θ. A random variable has a distribution that is an element of a family which we identify by a value of a parameter. This value may be known or unknown. Given that they index families of distributions in this way, we associate parameters with random variables and not with laws. The operator θ is used for this purpose, for example,

$$\theta(Y) = \lambda, \tag{2.2}$$

identifies λ as the parameter associated with Y.

2.2.4 DENSITIES AND LIKELIHOODS

For discrete and continuous random variables, probabilities are calculated through the use of probability density functions. If the random variable X is discrete, then the probability, $\Pr(X = x)$, is associated with a value of X and, if it is continuous, the probability, $\Pr[x < X < x + \mathrm{d}x)$ is associated with an interval $[X, X + \mathrm{d}x]$. In either case, the mathematical form of the density is unique for each law, while we also require a value of a random variable and a parameter for evaluation. Implementation reflects this structure; for example, the Poisson density is

$$\frac{\lambda^y}{e^\lambda y!}, \tag{2.3}$$

where y and λ are specific values of the random variable and parameter.

In general, a density is associated with the triple $\{\mathcal{L}, X, \theta\}$ but, when the uniqueness of a density for a given law is to be emphasized, it may be desirable to associate densities with laws only. While it may seem of little practical consequence to make such a distinction, it is important for the reader to realize that some computer algebra packages allow for this level of abstraction. Here we associate the pure mathematical form of a density with a law as $f(\mathcal{L})$ and the density for an element of the family given by $\mathcal{L}$ as $f(\mathcal{L})(X, \theta)$ the evaluation of $f(\mathcal{L})$ at the objects, (X, θ), that identify a particular element of the family. For example, the Poisson density is specified by

$$f(\text{Poisson}) = \frac{\exp\{-(\cdot_2)\}(\cdot_1)^{(\cdot_2)}}{(\cdot_1)!},$$

where the input $f(\text{Poisson})(y, \lambda)$ evaluates to the expression (2.3) and $\cdot_1, \cdot_2$ refer to the arguments y, λ, respectively. The subtle distinction is useful in studying model misspecifications, as we shall see in the examples that follow.

The mathematical form of a likelihood is identical to that of a density, up to a constant, and hence unique for a law. Likelihoods are then identified in a manner similar to that of densities. For density functions, the parameters are considered fixed and the value x is viewed as a variable. Interchanging these roles produces the likelihood function whose logarithm plays a central role in statistical inference. The log-likelihood is computed as the logarithm of the density associated with a particular law. The mathematical form of the likelihood is

$$l(\mathcal{L}) = \log(f(\mathcal{L})),$$

which may be evaluated by $l(\mathcal{L})(X, \theta(X))$. For example, for the case of the Poisson

$$l(\text{Poisson})(y, \lambda) \to y \log(\lambda) - \lambda - \log(y!).$$

2.2.5 GENERATING FUNCTIONS

Along with the densities and likelihoods, the moment generating functions are unique to laws. For a given law, their pure mathematical form is $\mathcal{M}(\mathcal{L})$, which may be evaluated for a specific distribution as $\mathcal{M}(\mathcal{L})(t, \theta)$. In the examples that follow, moment generating functions for laws are specified by assignment rather than being derived from the basic definition $E(\mathrm{e}^{Xt})$. For example, for the Poisson distribution, the moment generating is defined to be

$$\mathcal{M}(\text{Poisson}) = \exp\{(\cdot_2)(\exp\{(\cdot_1)\} - 1)\},$$

which is used below to compute the moment generating function for Y:

$$\mathcal{M}(\mathcal{L}(Y))(t, \theta(Y)) \to \mathrm{e}^{\lambda(\mathrm{e}^t - 1)}. \tag{2.4}$$

Now, when a moment generating function is available for a particular law, moments may be computed for a random variable with the same law. For the expected value operator we have the additional rule:

$$E(X^p) = \left\{ \frac{\partial^p}{\partial t^p} \mathcal{M}(\mathcal{L}(X))(t, \theta(X)) \right\} \Bigg|_{t=0}$$

Of course, in cases where a moment generating function does not exist, some other means of computing moments will be necessary.

The cumulant generating function for a specific law is defined as the natural logarithm of the moment generating function for that law:

$$\mathcal{K}(\mathcal{L}) = \log(\mathcal{M}(\mathcal{L}))$$

So, for the case of the Poisson we have

$$\mathcal{K}(\text{Poisson})(t, \lambda) \to \lambda(\mathrm{e}^t - 1).$$

2.3 Examples

In the examples that follow, we have defined Y, B, W, Z, and R to be random variables with the laws binomial, gamma, Gaussian, and hypergeometric. We also define the parameters, densities, and moment generating functions in each case. For example, for W this is accomplished by the following definitions:

$$f(\text{gamma}) = \Gamma(\cdot_2)^{-1}(\cdot_1)^{(\cdot_2)-1}\exp\{-(\cdot_1)\},$$

$$\mathcal{M}(\text{gamma}) = (1 - (\cdot_1))^{-(\cdot_2)},$$

$$\mathcal{L}(W) = \text{gamma},$$

$$\theta(W) = \alpha.$$

These are used below for evaluation:

$$f(\mathcal{L}(W))(W, \theta(W)) \to \Gamma(\alpha)^{-1}W^{\alpha-1}e^{-W},$$

$$\mathcal{M}(\mathcal{L}(W))(t, \theta(W)) \to (1-t)^{-\alpha}.$$

Note that we have constrained the law to identify only the standard gamma distribution with shape parameter α and not the larger family that includes a scalar parameter as well.

2.3.1 A FAMILIAR LIMIT RESULT

Showing that a Poisson process with rate λ gives rise to the Poisson distribution can be accomplished in a number of ways. One of these involves partitioning an interval of unit length into n equally spaced subintervals where a count on a subinterval can be effectively treated as a Bernoulli random variable with probability of success λ/n. The number of events in the entire interval, Y, then approximates a sum of Bernoulli's and hence is approximately binomial. The result is demonstrated by showing that the moment generating function for Y converges to that of the Poisson as n gets arbitrarily large:

$$\mathcal{L}(B) = \text{binomial},$$

$$\mathcal{M}(\text{binomial}) = (1 - \cdot_{2.2} + \cdot_{2.2}\,e^{\cdot_1})^{\cdot_{2.1}},$$

$$\theta(B) = \{n, \lambda/n\},$$

$$\lim_{n\to\infty}\mathcal{M}(\mathcal{L}(Y))(t, \theta(B)) \to e^{\lambda(e^{tx}-1)},$$

which is the same as (2.4). Most computer algebra packages have an operator for computing limits.

2.3.2 BARTLETT IDENTITIES

If the support of a distribution is independent of θ, then a set of identities due to M. S. Bartlett arises from the fact that densities integrate to one. If we successively differentiate $\int f_\theta(x)\,\mathrm{d}x = \int e^{l(\theta)}\,\mathrm{d}x = 1$ with respect to θ, and assume that integration and differentiation are interchangeable, a sequence of identities is obtained. The first of these gives the mean of the score function as 0 and the second gives its variance as the expected Fisher's information. These results can be verified for the Poisson distribution

$$u = \partial_\lambda l(\mathcal{L}(Y)(Y, \theta(Y))),$$

$$E(u) \to 0,$$

$$E(u^2) + E(\partial_\lambda u) \to 0,$$

where, in general, ∂_t denotes the operator $\partial/\partial t$.

2.3.3 MODEL MISSPECIFICATION

A model may be misspecified with our tools by mixing objects associated with different laws. For example, we might evaluate the likelihood for one law at a random variable that has a different law. The effect of such a misspecification could then be studied. We give an example of this below.

When a model is misspecified, Bartlett's identities will not necessarily hold. This is evident if one examines an alternative approach to showing that the expected score function is zero. Consider the integral $\int u(\theta)g(x)\,\mathrm{d}x$ where $u(\theta) = \partial_\theta \log\{f_\theta(x)\} = \{\partial_\theta f_\theta(x)\}/f_\theta(x)$ and f_θ is an assumed model. The true model for the data is represented by the density g. Now if g and f_θ coincide, then for the above integral we have

$$\int u(\theta)g(x)\,\mathrm{d}x = \int [\{\partial_\theta f_\theta(x)\}/f_\theta(x)]g(x)\,\mathrm{d}x$$

$$= \int [\{\partial_\theta f_\theta(x)\}/f_\theta(x)]f_\theta(x)\,\mathrm{d}x$$

$$= \int \{\partial_\theta f_\theta(x)\}\,\mathrm{d}x$$

$$= \partial_\theta \int f_\theta(x)\,\mathrm{d}x = \partial_\theta 1 = 0.$$

However, if $f \neq g$ then the above cancellation of f_θ and g will not occur and the identity may not obtain, similarly for other Bartlett identities. Nevertheless, the components of the Bartlett identities may still be computed.

Consider the case where the likelihood of a binomial random variable is approximated by the Poisson. Here the first Bartlett identity holds but the second does not:

$$\theta(Y) = \{n, \lambda/n\},$$

$$u = \partial_\lambda l(\mathcal{L}(X)(Y, \theta(X))),$$

$$E(u) \to 0,$$

$$E(u^2) + E(\partial_\lambda u) \to n^{-1}.$$

Note that the second identity will obtain asymptotically for reasons given in the first example. In general, when a model is misspecified this will not be the case. Consequently, standard statistics such as the Wald, score, and likelihood ratio are not asymptotically χ^2. This effect arises from improper standardization. In the case of the Wald statistic, this can be corrected by the use of sandwich estimators of asymptotic variance.

2.3.4 TAYLOR EXPANSION AND THE DELTA METHOD

Consider the case where we try to show that the first Bartlett identity obtains for the gamma distribution. Here the score function is given by

$$u(\alpha) = \partial_\alpha l(\mathcal{L}(W)(W, \theta(W))) \quad \rightarrow \quad \log(W) - \Psi(\alpha),$$

where Ψ denotes the digamma function. The difficulty here is that $\log(W)$ is not in a form for which expectations have been defined. E only knows how to handle polynomial functions of random variables. One way to fix this, and other cases like it, is to approximate $\log(W)$ by a polynomial through the use of a Taylor expansion. This is known by some as the delta method.

Most computer algebra packages have an operator for Taylor series expansion. For what follows, and for display purposes, we will denote this operator by S and assume that it returns an expansion of its argument truncated at three terms. This reflects standard practice when using the delta method to compute expected values rather than variances although expansions to much higher orders are feasible. Using S, we have

$$S(\log(W)) \rightarrow \log(\alpha) + \frac{(W - \alpha)}{\alpha} + \frac{(W - \alpha)^2}{2\alpha^2},$$

which is a form suitable for E.

Thus, for smooth non-polynomial functions g, the operator E requires the additional rule

$$E(g(X)) = E(S(g(X))).$$

Using this updated operator, we may now compute the expected value of the score function

$$E(\partial_\alpha l(\mathcal{L}(W)(W, \mathcal{L}(W)))) \rightarrow \frac{2\alpha \log(\alpha) - 1}{2\alpha^2} - \Psi(\alpha),$$

where the result is non-zero. The discrepancy is due to the truncation error of the delta method. A plot of the result given in Figure 2.1 shows that the approximation is useful at least for large values of α.

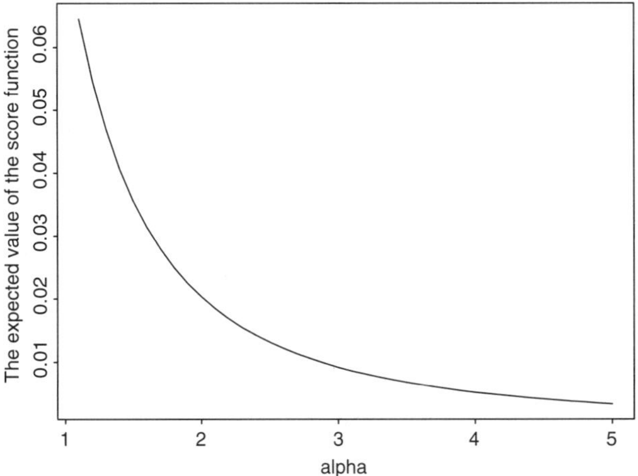

Fig. 2.1. The delta method works well for large values of α but not small values.

2.3.5 THE CAPTURE–RECAPTURE MODEL: AN EXAMPLE OF SYMBOLIC NEWTON–RAPHSON

In a scheme to estimate the size, P, of a wildlife population, k animals are captured, tagged, and released. After sufficient time has passed to allow the animals to re-integrate, n are recaptured. The number of animals, R, that are tagged in the sample may be used to conduct inference for P.

The random variable R has a hypergeometric distribution where

$$Pr(R = r) = p(r) = \frac{\binom{k}{r}\binom{P-k}{n-r}}{\binom{P}{n}}.$$

Plotting $p(r)$ as a function of P gives the likelihood for P, from which the maximum can be obtained. For example, suppose $k = 20$, $n = 10$, and $r = 4$, then, from Figure 2.2, $\hat{P}$ appears to be 50. However, computing $\hat{P}$ directly from the score function does not result in a closed-form expression in general. As a result, the solution is usually obtained numerically and one method for doing this automatically is the Newton–Raphson iteration. The purpose of this example is to show that in an equally automatic way an analytic approximation for $\hat{P}$ can be obtained using computer algebra, the result being that the moments and cumulants of the maximum likelihood estimate can be computed even when a closed-form expression does not exist.

In the same way that computer algebra packages have operators for Taylor series expansions, many have operators to give the inversion of a series. We denote as S^{-1} the operator that gives the truncated inverse series. Assuming that the law

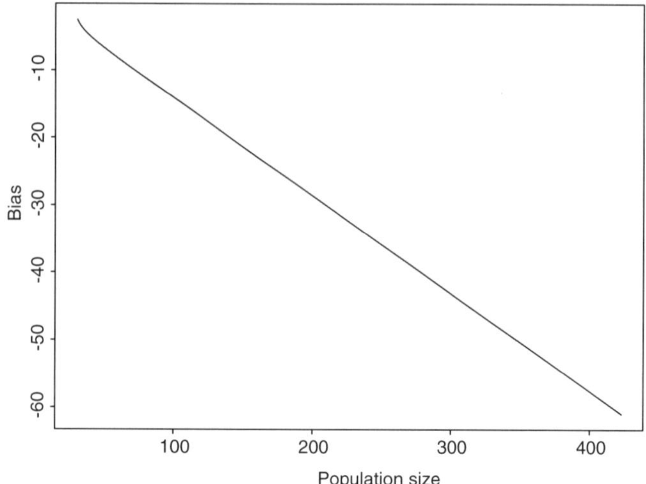

Fig. 2.2.   The log-likelihood function for the size of the animal population. From inspection the maximum likelihood estimate appears to be 50.

and density function have been defined, using S^{-1} we can derive an expression for $\hat{P}$:

$$\mathcal{L}(R) = \text{hypergeometric},$$

$$\theta(R) = P,$$

$$u(P) = \partial_P l(\mathcal{L}(R))(R, \theta(R)),$$

$$\hat{P} = S^{-1}(u(P))\big|_{P=0},$$

which yields an analytic expression for $\hat{P}$ in terms of R.

An expression for the expected value of $\hat{P}$ can now be found by simply applying the operator E to this expression. This will result in the repeated application of the delta method, as many Taylor series approximations are executed in this calculation. Assuming $k = 20$ and $n = 10$, we may compute the bias of $\hat{P}$ by $E(\hat{P}) - P$ and the result is plotted in Figure 2.3, which gives evidence of bias for the maximum likelihood estimate here.

2.3.6   A 'BARTLETT' CORRECTION

The use of likelihood ratio statistic, Λ, to test a hypothesis requires a reference distribution. A standard asymptotic result gives $-2\log(\Lambda) \approx \chi_p^2$ as a popular candidate. If this approximation is accurate then $E[-2\log(\Lambda)] \sim p$.

For the gamma distribution defined earlier, we can compute an analytic expression for $E[-2\log(\Lambda)]$, which should be 1 in this case. This is achieved through repeated Taylor series expansions that render $-2\log(\Lambda)$ as a polynomial in W.

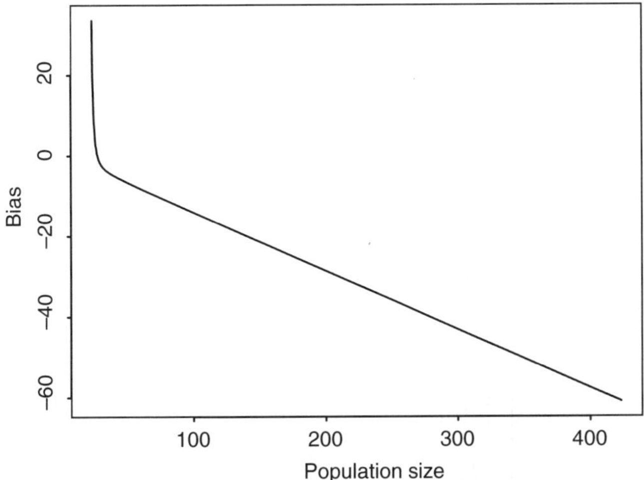

Fig. 2.3. The figure shows appreciable bias for the maximum likelihood estimate especially for small values of P. Note however that values of P less than 20 are inadmissible.

The desired expansion is derived via the steps:

(i)   compute a Taylor series of $-2\log(\Lambda)$ in $\hat{\alpha}$ about α;

(ii) replace $\hat{\alpha}$ with the analytic expression derived using symbolic Newton–Raphson; and

(iii) apply the operator E to the result.

These steps are accomplished by:

$$\hat{\alpha} = S^{-1}(u(\alpha))\big|_{\alpha=0},$$

$$lrs = 2(l(\mathcal{L}(W))(W, \hat{\alpha}) - l(\mathcal{L}(W))(W, \alpha)),$$

$$S(lrs) \rightarrow \frac{(-\log(W) + \Psi(\alpha))^2}{2\Psi'(\alpha)} - \frac{(-\log(W) + \Psi(\alpha))^4 \Psi''(\alpha)^2}{8\Psi'(\alpha)^5},$$

$$E(S(lrs))$$

$$\rightarrow \frac{1 - \log(\alpha) + \alpha \log(\alpha)^2 + \Psi(\alpha) - 2\alpha \log(\alpha)\Psi(\alpha) + \alpha\Psi(\alpha)^2}{2\alpha\Psi'(\alpha)}$$

$$+ \frac{(-\log(\alpha) + \Psi(\alpha))^2(-6 + 2\log(\alpha) - \alpha\log(\alpha)^2 - 2\Psi(\alpha))\Psi''(\alpha)^2}{8\alpha\Psi'(\alpha)^5}$$

$$+ \frac{(-\log(\alpha) + \Psi(\alpha))^2(2\alpha\log(\alpha)\Psi(\alpha) - \alpha\Psi(\alpha)^2)\Psi''(\alpha)^2}{8\alpha\Psi'(\alpha)^5}.$$

Of course, the final result does not equal 1 because the likelihood ratio is not chi-square in this case. However, Figure 2.4 shows that the approximation improves for large values of α.

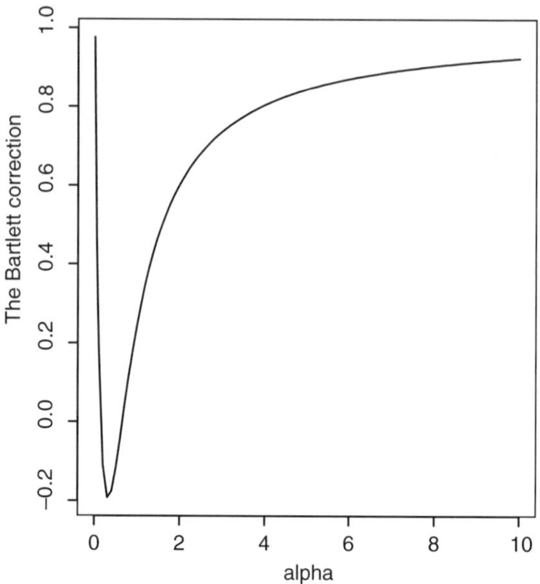

Fig. 2.4.   The Bartlett correction as a function of the shape parameter α.

2.3.7   SIMPLE LINEAR REGRESSION

In the last example, we had only one random variable; we had assumed a sample of size 1. This is obviously unrealistic and, typically in calculations of this type, the sample size is left unspecified; therefore, log-likelihoods are sums of arbitrary length. For the moment, we use the operator Σ to denote such sums. For example, consider simple linear regression through the origin:

$$f(\text{Gaussian}) = (2\pi \cdot_{2.2})^{-n/2} \exp\left\{ -\frac{(\cdot_1 - \cdot_{2.1})^2}{2 \cdot_{2.2}} \right\},$$

$$M(\text{Gaussian}) = \exp\left\{ \cdot_1 \cdot_{2.1} + \tfrac{1}{2}(\cdot_1)^2 \cdot_{2.2} \right\},$$

$$\mathcal{L}(Z) = \text{Gaussian},$$

$$\theta(Z) = \{\beta X, \tau\},$$

$$u(\beta) = \frac{\partial}{\partial \beta} l(\mathcal{L}(Z))(Z, \theta(Z)),$$

$$S^{-1}(u(\beta))\big|_{\beta=0} \rightarrow \Sigma(XY)/\Sigma(X^2).$$

Here the maximum likelihood estimate may be given explicitly. Further, X is not treated as a random variable, but as fixed, so the rule $E[\Sigma[\cdot]] = \Sigma[E[\cdot]]$ allows the computation of the first moment of $\hat{\beta}$. However, the computation of higher-order moments will require the computation of the higher-order moments of $\Sigma[YX]$. It is not immediately obvious how to do this and it is such computations that complicate matters to an extent that careful thought about the structure of the problem is required. Recognizing that this structure is common to many complex calculations leads to a broadly based kernel of operators that may be used in wide variety of problems. Such operators, and their application, constitute the remaining chapters of this text.

2.4 Remarks

The purpose of this chapter was to implement statistical concepts introduced in a first course in mathematical statistics. This has been accomplished with a very limited number of definitions. The rules for probability, random variables, and so forth, requires only about 30 statements each. In the following chapters, as with subsequent courses, generality is achieved through increased abstraction. In the following chapters, operators that manipulate sums of arbitrary length, vectors of parameters of arbitrary length, etc., will be developed.

2.5 Bibliographical notes

There is a growing kit of tools available for computer algebra calculations in mathematical statistics with libraries of operators accompanying most standard packages. Accompanying this is an increased use of such tools by statisticians in both research and teaching. Many books including Heller (1991), Davenport, Siret and Tournier (1988), and so forth, offer sound introductory material. They concentrate on problems in mathematical statistics one finds in an introductory course on the subject. This text is different, as such material is primarily confined to this chapter. Instead, we attempt to identify a kernel of symbolic operators that can be used in a wide variety of advanced problems one might find in graduate level courses or in research. Kendall (1993) provides a useful survey of computer algebra in probability and statistics.

Another implementation of basic probability is given by Pistone, Riccomagno and Wynn (1999, 2000). Here an indicator variable reflects whether or not an event has occurred. This allows set operators to be replaced by addition and multiplication. More complex events may be represented as polynomials rendering probability calculations in a manner well suited for a computer algebra package. Complicated problems can be handled by this approach; for example, Gröbner bases may be used to identify the design points in a large factorial experiment with confounded effects or to identify structural zeroes in a large contingency table.

2.6 Problems

The problems below illustrate further the application of the procedures defined in this chapter. They are intended to be both challenging and instructive.

1. Implement and demonstrate De Morgan's laws

$$\left(\cup_{i=1}^{k} A_i\right)^{c} = \cap_{i=1}^{k} A_i^{c}$$
$$\left(\cap_{i=1}^{k} A_i\right)^{c} = \cup_{i=1}^{k} A_i^{c}$$

for sets $A_1, \ldots, A_k$.

2. Define indicator functions for sets and use addition and multiplication to implement set functions and a probability measure. Show that these functions can be used to perform some of the calculations done in this chapter.

3. Suppose a disease is lethal for individuals of genotype $g[a, a]$, in the notation of the population genetics example. Use this fact to compute the conditional distribution of the genotypes of surviving offspring and investigate the Hardy–Weinberg equilibrium by calculating the distribution of genotypes of children in successive generations. Does the allele a disappear or does the population reach equilibrium? How many generations does it take? Is the distribution just the original distribution conditioned on survival?

4. Investigate the above problem when the disease, like Duchenne muscular dystrophy, affects only males.

5. Extend the population genetics example to include two children. Use this to obtain the distribution of the genotypes of the two children. Now, if a represents the allele of a recessive disease, only the genotype $g[a, a]$ is affected. Using this fact, replace each pair of genotypes with a function which counts the number of affected persons to find the distribution of this number.

   If a mother has two affected children, what is the probability that she is affected?

6. The probability generating function of a random variable X is defined to be $G[z] = E[z^{X}]$. It may be computed from the generating function $\mathcal{M}(t)$ by setting $t = \log(z)$. Compute successive derivatives $G^{(i)}(z)$ and evaluate these at $z \to 0$ for the binomial and the Poisson distributions. Verify the relation between these derivatives and the probabilities of $X = 0, X = 1, \ldots$ for these distributions.

7. Devise a method for identifying independent variables. If X and Y are independent, the generating function of $T = X + Y$ is the product of the generating functions of the components. Use this fact together with the previous exercise to compute the first few probabilities of the sum of a binomial and an independent Poisson random variable.

8. Write a function to compute correlations. Use this function to compute the correlation between the deviation $X - E[X]$ and its square, $(X - E[X])^2$, for the Gaussian, Poisson, and binomial distributions. When is this correlation zero? When is it independent of the parameters of the distribution?

9. Approximate the moments of $X Y^{-1/2}$ assuming that X and Y are independent.

The standard t-statistic has this form where X is Gaussian and Y is independent and proportional to a χ^2_{n-1} distribution which has moment generating function $(1 - 2t)^{-(n-1)/2}$. Use the result to compute approximations for the first few moments of the t-distribution. Of course, the moments do not always exist. When and why is this approximation working?

10. Calculate, in your head, the moment generating function of the Bernoulli distribution. The moment generating function of the sum of n independent random variables with some distribution is just the moment generating function of the distribution raised to the power n. Extend the rules for laws and moment generating functions to incorporate this fact and apply it to the generating function you just did in your head, thereby deriving the moment generating function of the binomial distribution.

11. Define X and Y to have a bivariate Gaussian distribution with mean $\{\mu_X, \mu_Y\}$, and correlation ρ. Compute the variance of $(X - \mu_X)(Y - \mu_Y)$.

12. Cumulants κ_i are derivatives of the cumulant generating function, $\mathcal{K}(t) = \log\{\mathcal{M}(t)\}$ evaluated at $t \to 0$. Expressing these derivatives as cumulants, write a procedure to compute the Taylor expansion of $\mathcal{K}(t)$ about $t = 0$. Using the relation defining the cumulant generating function, compute the related moment generating function. By differentiating this function, derive the expression of the first five moments in terms of cumulants.

13. By interchanging the roles of moment and cumulant generating functions in the above problem, derive the expression of the first five cumulants in terms of moments.

14. Define procedures to convert moments to cumulants and from cumulants to moments. Verify that these are inverses by converting $E[X^3]$ to cumulants and convert these cumulants back to moments.

15. Using the definition of the cumulants, show, by hand, that cumulants have the useful property that if $T = \sum_{i=1}^{n} X_i$, where the X_i are independent random variables, the cumulants of T are n times the cumulants of X_i. Use the relation of cumulants and moments to derive expressions for the expectation $E[T^2]$ in terms of $E[X_i^j]$.

16. For IID random variables, the expectation of a product of sums may be expressed as a product of moments. For example,

$$E\left[\left(\sum_i X_i\right)^2\right] = nE[X_i^2] + n(n - 1)(E[X_i])^2.$$

These sums are unbiased estimates of the corresponding linear combination of the product of moments. Compute the coefficients of this linear relation relating moments to sums for products of j sums, $1 \le j \le 3$, to create the system of linear equations $E[(\sum X)^j] = MV$, where M is a matrix of coefficients and V is a vector of products of expectations, $E[X^j]$.

By solving this system of linear equations, find unbiased estimates of $E[X^j]$ for $1 \le j \le 3$.

17. Use the symbolic Newton–Raphson technique to compute the first two moments of $\hat{\alpha}$ for the gamma distribution. Use properties of the digamma function to compute expected values.

18. Show that the first two Bartlett identities hold for the hypergeometric distribution considered in this chapter.

19. Compute the moments for $\hat{\psi}$ for the Weibull distribution. Use facts about the digamma function to compute expected values.

20. Compute the Bartlett correction for the likelihood ratio statistic for the Weibull distribution.

3
Fundamental procedures

3.1 Introduction

We concluded Chapter 2 pondering how the expected value operator should be extended for computations involving sums of unspecified length. This question leads to a fundamental algorithm for the full partition of a set which will be useful for this and many other calculations.

Consider the sum $Y = \sum_{i=1}^{n} X_i$. If n is not specified, a symbolic representation of Y will not have the explicit form $X_1 + X_2 + \cdots + X_n$, but will remain unexpanded. This means that for our current version of E the distributive law will not be executed unless we add the rule: $E(\sum_i X_i) = \sum_i E(X_i)$. However, this extension of the operator does not solve the problem of how to compute the second-, third- or higher-order moments of Y. For example, to compute the second moment, $E[Y^2]$, one typically expands Y^2 into two terms

$$Y^2 = \sum_i X_i^2 + \sum_{i \neq j} X_i X_j,$$

since the behaviour of E will differ for each term. For instance, if Y is a sum of independent, random variables then the expected value operator will factor for the second term, yielding

$$E(Y^2) = \sum_i E(X_i^2) + \sum_{i \neq j} E(X_i)E(X_j)$$

$$= \sum_i E(X_i^2) + \left\{ \sum_i E(X_i) \right\}^2 - \sum_i E(X_i)^2.$$

To elucidate developments in this chapter, consider computing the expected value of the product $\Pi = \sum X_i \sum Y_i \sum Z_i$. Π can be expanded to yield

$$\sum_i X_i Y_i Z_i + \sum_{i \neq j} X_i Y_i Z_j + \sum_{i \neq j} X_i Y_j Z_i + \sum_{i \neq j} X_j Y_i Z_i + \sum_{i \neq j \neq k} X_i Y_j Z_k. \tag{3.1}$$

where an object like $\sum_{i \neq j \neq k} X_i Y_j Z_k$ is referred to as a disjoint sum. If the operator E is to be extended to objects like Π, then we must automate the expansion of such objects.

Each term in the above expansion can be identified by a distinct partition of the set $S = \{X, Y, Z\}$. The blocks of each partition are themselves identified by the subscripts. That is, symbols with the same subscript belong to the same block. For example, the disjoint sum $\sum_{i \neq j} X_i Y_j Z_i$ is identified by the partition $(XZ|Y)$. The entire expansion, a collection of five terms, is determined by the collection of all the possible partitions of S, namely,

$$\mathcal{P}[S] = \{(XYZ), (Z|XY), (Y|XZ), (X|YZ), (X|Y|Z)\}.$$

Hence, automating the expansion of Π may be done by computing all partitions of S.

In this chapter, the collection of all the partitions of the set S is called the full partition of S and is denoted as $\mathcal{P}[S]$. The notation $\mathcal{P}[S]$ also represents the result of applying the transformation $\mathcal{P}$ to the set S. This transformation is central to our discussion. It provides the basis for automating a large class of calculations and its use is by no means restricted to calculations that involve sums. Writing moments in terms of cumulants, cumulants in terms of moments, sample moments in terms of k-statistics, or generalized cumulants in terms of ordinary cumulants are just a few examples of calculations that are expressed using full partitions. In fact, any result associated with the inclusion–exclusion rule that defines the full partition may be computed using the algorithm $\mathcal{P}$.

Exploiting the common structure of seemingly different calculations is one theme of this chapter. Doing so leads to the creation of a computer algebra involving lists, ω, to represent objects like Π, transformations of lists, $\mathcal{T}$, to automate calculations like the expansion of Π, and functions, f, that translate between objects and lists. Basic transformations automate all calculations and are derived from identities like those of Section 3.10, identities that themselves can be written in terms of five basic functions and four basic transformations. The process of moving from identities to complex transformations in a proof is initially expressed in a traditional way, but is ultimately automated. The latter, which comprises a computer algebra proof, requires the exploration of the properties of our algebra of operators $(\omega, f, \mathcal{T})$.

3.2 Partitions of a set

Consider a set of k objects $V_k = \{X^1, \ldots, X^k\}$. A partition $p = (b_1| \cdots |b_m)$ of V_k divides V_k into mutually exclusive and exhaustive sets, or blocks, b_j, where order within a block does not matter. For example, $(XZ|Y)$ is a partition of $\{X, Y, Z\}$ with two blocks: $b_1 = XZ$, and $b_2 = Y$. The set of all such partitions is called the full partition of V_k, denoted $\mathcal{P}[V_k]$.

A full partition can be computed by iterating an inclusion–exclusion rule. For each element, $p = (b_1| \cdots |b_m) \in \mathcal{P}[V_k]$, elements of $\mathcal{P}[V_{k+1}]$ are generated by adding a new object X_{k+1} to each block of p (inclusion) and by adding X_{k+1} as a separate block (exclusion). For example, the full partition of $\{X, Y, Z\}$ can be constructed iteratively by the steps:

$$\mathcal{P}[\{X\}] = \{(X)\},$$

$$\mathcal{P}[\{X, Y\}] = \{(XY), (X|Y)\},$$

$$\mathcal{P}[\{X, Y, Z\}] = \{(XYZ), (XY|Z), (XZ|Y), (X|YZ), (X|Y|Z)\}.$$

In the process, the elements (XY), (XYZ), $(XZ|Y)$, $(X|YZ)$ result from the inclusion part of the rule, while the elements $(X|Y)$, $(XY|Z)$, $(X|Y|Z)$ result from the exclusion part.

3.3 Some important uses of $\mathcal{P}$

Many calculations implicitly involve the inclusion–exclusion rule and hence may be automated using $\mathcal{P}$. We demonstrate this for some important cases used throughout the rest of the book. In each case, we first identify the use of the inclusion–exclusion rule and this allows us to immediately state a general identity in terms of a full partition. Throughout, a typical element p of a full partition is always assumed to have the form $p = (b_1|\cdots|b_m)$. In addition, for X^j and t^j the use of a superscript denotes a distinct variable, indexed by $j = 1, \ldots, k$, and not a power.

3.3.1 EXPANDING A PRODUCT OF SUMS

Expanding the product $\sum X_i \sum Y_i \sum Z_i$, (3.1), will, at some point, require expanding the product of $\sum_{i \neq j} X_i Y_j$, a disjoint sum identified by the partition $(X|Y)$, with the sum $\sum Z_k$. The result

$$\sum_{i \neq j} X_i Y_j Z_i + \sum_{i \neq j} X_i Y_j Z_j + \sum_{i \neq j \neq k} X_i Y_j Z_k$$

can be identified by the set of partitions $\{(XZ|Y), (X|YZ), (X|Y|Z)\}$ obtained by applying the inclusion–exclusion rule to $(X|Y)$. The first two terms result from the inclusion part of the rule and the final term from the exclusion part. That is, the rule used to expand a product of sums is the *same* inclusion–exclusion rule used by $\mathcal{P}$.

From this we can immediately obtain a general rule for expanding a product of k sums

$$\prod_{j=1}^{k} \sum_{i=1}^{n} X_i^j = \sum_{p \in \mathcal{P}[V_k]} \sum_{i_1 \neq \cdots \neq i_m} \chi_{i_1}^{b_1} \cdots \chi_{i_m}^{b_m}, \tag{3.2}$$

where $\chi_i^b = \prod_{j \in b} X_i^j$.

3.3.2 EVALUATING A DISJOINT SUM

Since expanding a product of sums depends on the full partition of a set, so will the inverse operation: expressing a disjoint sum in terms of sums of products of sums. Consider expressing the sum $\sum_{i \neq j \neq k} X_i Y_j Z_k$ as a product of sums. At some point, this will require mapping $\sum_i \sum_j \sum_{k \neq \{i,j\}} X_i Y_j Z_k$ to

$$\sum_i X_i \sum_j Y_j \sum_k Z_k - \sum_i X_i Z_i \sum_j Y_j - \sum_i X_i \sum_j Y_j Z_j.$$

The result can be identified by the set of partitions $\{(X|Y|Z), (XZ|Y), (X|YZ)\}$ obtained by applying the inclusion–exclusion rule to $(X|Y)$, a partition that identifies the term $\sum_i X_i \sum_j Y_j$. Here the first term results from the exclusion part of the rule and the final two terms from the inclusion part. That is, the rule used is the *same* inclusion–exclusion rule used by $\mathcal{P}$.

The general rule for disjoint sums is

$$\sum_{i_1 \neq \cdots \neq i_k} X_{i_1}^1 \cdots X_{i_k}^k = \sum_{p \in \mathcal{P}[V_k]} c^p \prod_{b \in p} \sum_{i=1}^n \chi_i^b, \tag{3.3}$$

where $\chi_i^b = \prod_{j \in b} X_i^j$ as in the last example and

$$c^p = (-1)^{k-m} \prod_{j-1}^m (|b_j| - 1)!.$$

Here $|b|$ denotes the number of elements in the set b.

3.3.3 MOMENTS $\longleftrightarrow$ CUMULANTS

The joint moment generating function for X, Y and Z is denoted by M_S and is defined as

$$M_S(t) = E[e^{S \cdot t}],$$

where $S = \{X, Y, Z\}$ and $t = \{t_X, t_Y, t_Z\}$. The cumulant generating function, denoted by K_S, is defined as

$$K_S(t) = \log\{M_S(t)\}.$$

Using these we may find joint moments and cumulants, denoted by μ and κ, respectively, as follows:

$$\mu_{XYZ} = E[XYZ] = \{\partial_{XYZ} M_S(t)\}|_{t=0},$$

$$\kappa_{XYZ} = \{\partial_{XYZ} K_S(t)\}|_{t=0},$$

where $\partial_X = \partial/\partial t_X$, $\partial_{XY} = \partial^2/\partial t_X \partial t_Y$ and so forth. Now by virtue of the identity

$$M_S(t) = \exp\{K_S(t)\},$$

we may write moments in terms of cumulants or *vice versa* through repeated differentiation. For example, suppressing the arguments of $M_S(t)$ and $K_S(t)$, we have the relation

$$\partial_X M = e^K \partial_X K$$

giving $\mu_X = \kappa_X$. Now successive differentiation with respect to t_Y requires the product rule for differentiation, which mimics the inclusion–exclusion rule

$$\partial_{XY} M = e^K \{\partial_{XY} K + \partial_X K \times \partial_Y K\}.$$

The result is a sum indexed by the partitions $\{(XY), (X|Y)\}$ which result from applying the inclusion–exclusion rule to (X), the partition that identifies $e^K \partial_X K$. Setting $t = 0$ gives moments in terms of cumulants

$$\mu_X = \kappa_X,$$

$$\mu_{XY} = \kappa_{X,Y} + \kappa_X \kappa_Y.$$

The general rule for writing moments in terms of cumulants for a set of random variables V_k is

$$\mu_{V_k} = \sum_{p \in \mathcal{P}[V_k]} \prod_{b \in p} \kappa_b, \tag{3.4}$$

where $\kappa_b = \{[\prod_{j \in b} \partial_{t^j}] K_{V_k}(t)\}_{|t=0}$ is the joint cumulant for the components of the block b.

Since writing moments in terms of cumulants depends on the full partition of a set, so will the inverse operation: writing cumulants in terms of moments. The general rule for this is

$$\kappa_{V_k} = \sum_{p \in \mathcal{P}[V_k]} c_p \prod_{b \in p} \mu_b, \tag{3.5}$$

where the coefficient $c_p = (-1)^{m-1} \times (m-1)!$.

This last example is a case of the exlog relations, which assert that for two power series $B(t)$ and $C(t)$, where $B(t) = \exp\{C(t)\}$ and hence $C(t) = \log\{B(t)\}$, the coefficients in the expansion of one series may be written in terms of the coefficients in the expansion of the other. The coefficients are related as sums over a full partition.

3.4 A fundamental algorithm

The remainder of this chapter concerns operators that may be implemented directly in a computer algebra environment. The full partition algorithm, $\mathcal{P}$, will be applied in this book to lists whose components represent sums, moments, cumulants, and so forth. These components are linear in their arguments and hence we define $\mathcal{P}$ to be linear in its arguments:

$$\mathcal{P}[aA + bB + \cdots] = a\mathcal{P}[A] + b\mathcal{P}[B] + \cdots,$$

where $A, B, \ldots$ are sets and $a, b, \ldots$ are scalars. The argument of $\mathcal{P}$ is an unordered (sorted) set. Later, in applications where multiplication does not commute, the algorithm will be applied to ordered sets, but no modification is required of the transformation, only of the set.

The inclusion–exclusion rules may be implemented recursively as follows:

$$\mathcal{P}[\{X\}] = \{\{X\}\},$$

$$\mathcal{P}[\{X, Y, \ldots\}] = \mathcal{Q}_X(\mathcal{P}[\{Y, \ldots\}]),$$

where $\mathcal{Q}$ is defined as a linear operator that automates the inclusion–exclusion rule via the function $\mathcal{I}$,

$$\mathcal{Q}_X(\{A, B, \ldots\}) = \mathcal{I}_{\{X\}}(\{A, B, \ldots\}) + \{\mathcal{I}_X(A), B, \ldots\} +$$

$$\{A, \mathcal{I}_X(B), \ldots\} + \cdots.$$

and $\mathcal{I}$ is defined by

$$\mathcal{I}_X(A) = \{X\} \vee A,$$

where $\vee$ denotes 'join' and not 'union'. That is, $\{X\} \vee \{X\}$ gives $\{X, X\}$ and not $\{X\}$.

3.5   Objects, types, lists, functions, and transformations

In computer algebra, the basic unit of computation is a list. Although expressions are displayed in a standard fashion, their internal representation is as a list. We capitalize on this by representing objects, such as products of sums of expectations, as lists denoted by ω. All objects can be represented by a list with the same structure. The same list can then represent different objects. To avoid confusion the notion of a 'type' of object, or list, is introduced. This is similar to a sequence of 0's and 1's having multiple meanings such as character, integer, and so forth, the unique meanings being distinguished by types. Objects represented by lists ultimately will need to be represented by standard mathematical expressions. Functions, denoted f, are defined for this purpose.

Calculations are implemented as transformations $\mathcal{T}$ of lists of one type to linear combinations of lists of another type. The process of deriving transformations from identities is automated in Section 3.11 in a manner similar in spirit to deriving the law of total probability from identities for unions and mutually exclusive intersections. The form of each transformation is justified by computer algebra proof. However, we initially identify transformations in a traditional way to ease the reader into the realm of computer algebra proof.

The reader needs to come to terms with three ideas: how an object is represented as a list, how functions invert this process, and how identities lead to transformations of lists. Beginning with a list, any calculation is automated as

$$f \circ \mathcal{T} \circ \omega = f[\mathcal{T}[\omega]],$$

from which the meaning of '$\circ$' is clear. Our task is to identify ω, f, $\mathcal{T}$ for a number of calculations.

The components ω, f, $\mathcal{T}$ will be defined to be linear operators or, in the case of multiple arguments, multi-linear operators. For example, ω is defined as multilinear by assigning to lists the following rules:

$$\{\ldots, X + Y, \ldots\} = \{\ldots, X, \ldots\} + \{\ldots, Y, \ldots\},$$

$$\{\ldots, a\{\ldots\}, \ldots\} = a\{\ldots, \{\ldots\}, \ldots\}.$$

Similarly, any function or transformation, g, is given the rule

$$g \circ (a\{\ldots\} + \{\ldots\}) = a(g \circ \{\ldots\}) + g \circ \{\ldots\}.$$

There are many advantages to defining an algebra with this structure. Using a list as a common data structure for objects allows transformations to be composed easily. The output of one transformation is always suitable as input for another. The list representation avoids the notational baggage of a particular object which can be inefficient. For example, when computing the expected value of a product of sums, there is never any need to generate objects written explicitly as disjoint sums, as this is an intermediate step and the final result will ultimately be written in terms of sums, not disjoint sums. Avoiding explicit representations for intermediate steps boosts efficiency. Finally, representing all objects as lists allows us to capture the common structure of seemingly different calculations, such as expanding means and writing moments in terms of cumulants.

3.5.1 A COMMON DATA STRUCTURE

Consider the product of sums of products

$$t_1 = \sum XY \sum Z.$$

This object has a nested structure of the form

$$\Pi(\Sigma[\Pi(\cdots)]).$$

We may represent t_1 as a list by replacing each operator with a 'list' operator:

$$t_1 \equiv \{\{\{X, Y\}\}, \{\{Z\}\}\}.$$

Now only sums ever have a single argument and using a list to represent a sum seems redundant. Therefore, the list representation can be simplified to

$$t_1 \equiv \{\{X, Y\}, \{Z\}\}.$$

Similarly, the product of means of products $t_2 = E[XY]E[Z]$ has the nested structure

$$\Pi(E[\Pi(\cdots)])$$

and can represented by the same list,

$$t_2 \equiv \{\{X, Y\}, \{Z\}\},$$

where again the use of a list operator to represent expectation would be redundant.

Table 3.1. Standard objects

Type	$f_{\text{Type}} \circ \omega$
Σ	$\sum XY \sum Z$
Δ	$\sum_{i \neq j} X_i Y_i Z_j$
μ	$E[XY]E[Z]$
κ	$\kappa_{X,Y}\kappa_Z$
$\Sigma[\mu]$	$\sum E[X_i Y_i] \sum E[Z_i]$
$\Delta[\mu]$	$\sum_{i \neq j} E[X_i Y_i]E[Z_j]$
$\Sigma[\kappa]$	$\sum \kappa_{X_i,Y_i} \sum \kappa_{Z_i}$
$\Delta[\kappa]$	$\sum_{i \neq j} (\kappa_{X_i,Y_i}\kappa_{Z_j})$

Types are used to distinguish when $\omega = \{\{X, Y\}, \{Z\}\}$ represents t_1 rather than t_2 or any other object. The type for t_1 is denoted by Σ for sums. We define eight standard types representing different objects. There is a type to represent sums of moments, a type for sums of cumulants etc. These eight types form a basis for defining transformations and correspond to the vertices of Figure 3.1. They are given in Table 3.1 along with the mathematical representation of the list ω for that type. The next section gives details of such functions.

We eventually move to a more complicated list structure. However, for the moment, the current structure serves our purposes well as it simplifies initial developments. In addition, the use of sums rather than averages eases the exposition. We move to averages later.

3.5.2 FUNCTIONS OF LISTS

The final result of a calculation can be written explicitly in a familiar form by the application of a function f to a list. The function is a linear operator and is constructed by the composition of the operators given in Table 3.2.

For any of these functions, the presence of a subscript denotes the level of the list to which the operator is applied. For example,

$$\pi_2[\{\{X, Y\}, \{Z\}\}] \longrightarrow \{XY, Z\},$$

$$\Sigma_2[\{XY, Z\}] \longrightarrow \left\{\sum XY, \sum Z\right\},$$

$$\pi_1\left[\left\{\sum XY, \sum Z\right\}\right] \longrightarrow \sum XY \sum Z,$$

where, for clarity, the subscripts on sums are suppressed. Thus, $\sum XY$ denotes $\sum_{i=1}^{n} X_i Y_i$. A function for a type is constructed by composing these elementary functions. For example, $f_\Sigma = \pi_1 \circ \Sigma_2 \circ \pi_2$ applied to $\{\{X, Y\}, \{Z\}\}$ yields $\sum XY \sum Z$.

The sequences of functions to produce the standard objects are given in the second column of Table 3.3. The application of these functions to an object of that type represented by $\omega = \{\{X, Y\}, \{Z\}\}$ are given in column 3.

Table 3.2. Standard functions

Function	Example
Σ	$\Sigma \circ X \longrightarrow \Sigma X$
Δ	$\Delta \circ \{X, Y\} \longrightarrow \Sigma_{\neq}[X, Y]$
μ	$\mu \circ X \longrightarrow \mu_X = E[X]$
Π	$\Pi \circ \{X, Y\} \longrightarrow XY$
κ	$\kappa \circ \{X, Y\} \longrightarrow \kappa_{X,Y}$

Table 3.3. Standard objects

Type	Function f_{Type}	$f_{\text{Type}} \circ \omega$
Σ	$\pi_1 \circ \Sigma_2 \circ \pi_2$	$\sum XY \sum Z$
Δ	$\Delta_1 \circ \pi_2$	$\sum_{i \neq j} X_i Y_i Z_j$
μ	$\pi_1 \circ \mu_2 \circ \pi_2$	$E[XY]E[Z]$
κ	$\pi_1 \circ \kappa_2$	$\kappa_{X,Y}\kappa_Z$
$\Sigma[\mu]$	$\pi_1 \circ \Sigma_2 \circ \mu_2 \circ \pi_2$	$\sum E[X_i Y_i] \sum E[Z_i]$
$\Delta[\mu]$	$\Delta_1 \circ \mu_1 \circ \pi_2$	$\sum_{i \neq j} E[X_i Y_i]E[Z_j]$
$\Sigma[\kappa]$	$\pi_1 \circ \Sigma_2 \circ \kappa_2$	$\sum \kappa_{X_i,Y_i} \sum \kappa_{Z_i}$
$\Delta[\kappa]$	$\Delta_1 \circ \kappa_2$	$\sum_{i \neq j} (\kappa_{X_i,Y_i} \kappa_{Z_j})$

3.5.3 TRANSFORMATIONS

A small number of fundamental identities relate objects of different types. As a simple example, the identity

$$E\left[\sum X_i\right] = \sum E[X_i]$$

equates the expectation of a sum with a sum of expectations. The identities of Section 3.3 provide further examples. This section concerns the computational aspects of the implementation of such identities presented in a traditional way. Identities induce transformations. Examination of the implementation of an identity yields a transformation. The identities and their use to automatically derive transformations will be set out more generally in Sections 3.9–3.11.

Calculations such as computing moments from cumulants, or the expected value of a product of sums, are implemented as transformations of lists. Lists of one type may be expressed in terms of lists of another type. The re-expression is a transformation of the first type of list into a linear combination of lists of the second type. Because transformations are defined to be linear and produce results of a known form, they may be composed to permit more complex calculations.

A transformation of a list of type B into one of type A is denoted $\mathcal{T}_{A,B}$. Hence, writing an object of type B in terms of those of type A is accomplished by virtue of the relationship

$$f_B \equiv f_A \circ \mathcal{T}_{A,B}.$$

These transformations are constructed from four basic linear operators $\mathcal{P}$, $\mathcal{F}$, $\mathcal{C}$, and $\mathcal{D}$ for lists. Hence, $\mathcal{T}_{A,B}$ is also linear. $\mathcal{P}$ is the full partition operator, while the operators $\mathcal{F}$ and $\mathcal{D}$ flatten and distribute structure as

$$\mathcal{F}[\{\ldots, \{\ldots, X, \ldots\}, \ldots, \{\ldots, Y, \ldots\}, \ldots\}] \longrightarrow \{\ldots, X, \ldots, Y, \ldots\},$$
$$\mathcal{D}[\{\ldots\{\ldots, X, \ldots, Y, \ldots\}\ldots\}] \longrightarrow \{\ldots, \{X\}, \ldots, \{Y\}, \ldots\}.$$

$\mathcal{C}$ is a coefficient operator that acts as

$$\mathcal{C}^\bullet \circ \omega = c\,\omega,$$

where the value of c depends on the superscript of $\mathcal{C}$.

For any of these operators, the presence of a subscript denotes the level of the list to which the operator is applied. For example,

$$\mathcal{P}_2[\{\{X, Y\}, \{Z\}\}] \longrightarrow \{\mathcal{P}[\{X, Y\}], \mathcal{P}[\{Z\}]\}.$$

A specific transformation is specified as $\mathcal{T}_{A,B} = \mathcal{D}_2 \circ \mathcal{F}_2 \circ \mathcal{P}_1 \circ \cdots$. The next section gives examples of deriving transformations informally from identities.

3.6 Examples

For what follows, the reader must constantly bear in mind the list representation of objects and grow increasingly accustomed to the action of transformations on lists. For example, the expansion of the product of sums, $\sum XY \sum Z$ (type Σ) is automated by the application of $\mathcal{P}_1$ to its list representation $\{\{X, Y\}, \{Z\}\}$. Such exercises are repeated constantly in the following examples and are essential for the reader's comprehension. However, these exercises are ultimately redundant as the process of deriving transformations from identities is automated in Sections 3.9–3.11.

3.6.1 $\mathcal{T}_{\Delta,\Sigma}$: SUMS TO DISJOINT SUMS

An example of identity (3.2) is

$$\sum_i X_i Y_i \sum_i Z_i = \sum_i X_i Y_i Z_i + \sum_{i \neq j} X_i Y_i Z_j.$$

We may write this in terms of our operators. Table 3.1 is helpful for this purpose:

$$\Pi_1 \circ \Sigma_2 \circ \Pi_2 \circ \{\{X, Y\}, \{Z\}\}$$

$$= \sum_i X_i Y_i \sum_i Z_i$$

$$= \sum_i X_i Y_i Z_i + \sum_{i \neq j} X_i Y_i Z_j$$

$$= \Delta_1 \circ \Pi_2 \circ \{\{X, Y, Z\}\} + \Delta_1 \circ \Pi_2 \circ \{\{X, Y\}, \{Z\}\}$$

$$= \Delta_1 \circ \Pi_2 \circ (\{\{X, Y, Z\}\} + \{\{X, Y\}, \{Z\}\})$$

$$= \Delta_1 \circ \Pi_2 \circ \mathcal{F}_2 \circ (\{\{\{X, Y\}, \{Z\}\}\} + \{\{\{X, Y\}\}, \{\{Z\}\}\})$$

$$= \Delta_1 \circ \Pi_2 \circ \mathcal{F}_2 \circ \mathcal{P}_1 \circ \{\{X, Y\}, \{Z\}\}.$$

Now if we examine the two extremes of this expression, and drop the list $\{\{X, Y\}, \{Z\}\}$ in each case, we arrive at an identity, written in terms of our operators, that is equivalent to (3.2), i.e.

$$\Pi_1 \circ \Sigma_2 \circ \Pi_2 \equiv \Delta_1 \circ \Pi_2 \circ \mathcal{F}_2 \circ \mathcal{P}_1.$$

Similarly, if we examine the two extremes, and drop all the functions in each case, we obtain a transformation of a list of type Σ into lists of type Δ. That is, we identify $\mathcal{T}_{\Delta, \Sigma}$:

$$\mathcal{T}_{\Delta, \Sigma} \equiv \mathcal{F}_2 \circ \mathcal{P}_1.$$

It is in the above manner that we derive the remaining identities of Section 3.3 and their induced transformations.

3.6.2 $\mathcal{T}_{\Sigma, \Delta}$: DISJOINT SUMS TO SUMS

An example of (3.3) is

$$\sum_{i \neq j} X_i Y_i Z_j = \sum_i X_i Y_i \sum_i Z_i - \sum_i X_i Y_i Z_i.$$

Writing this in terms of our operators gives

$$\Delta_1 \circ \Sigma_2 \circ \{\{X, Y\}, \{Z\}\}$$

$$= \sum_{i \neq j} X_i Y_i Z_j$$

$$= \sum_i X_i Y_i \sum_i Z_i - \sum_i X_i Y_i Z_i$$

$$= \Pi_1 \circ \Sigma_2 \circ (\{\{X, Y\}, \{Z\}\} - \{\{X, Y, Z\}\})$$

$$= \Pi_1 \circ \Sigma_2 \circ \mathcal{F}_2 \circ (\{\{\{X, Y\}\}, \{\{Z\}\}\} - \{\{\{X, Y\}, \{Z\}\}\})$$

$$= \Pi_1 \circ \Sigma_2 \circ \mathcal{F}_2 \circ \mathcal{C}_2^p \circ (\{\{\{X, Y\}\}, \{\{Z\}\}\} + \{\{\{X, Y\}, \{Z\}\}\})$$

$$= \Pi_1 \circ \Sigma_2 \circ \mathcal{F}_2 \circ \mathcal{C}_2^p \circ \mathcal{P}_1 \circ \{\{X, Y\}, \{Z\}\},$$

where

$$\mathcal{C}_i^p \circ \omega = ((-1)^{l_i \circ \mathcal{F}_{i-1} \circ \omega - l_1 \circ \omega} \Pi_{i-1} \circ l_i \circ \omega)\omega \tag{3.6}$$

is just the coefficient of (3.3) written in terms of our operators. Here $l \circ \omega$ gives the length of the list ω.

As in the previous example, we can use the above expression to arrive at an identity

$$\Delta_1 \circ \Sigma_2 \equiv \Pi_1 \circ \Sigma_2 \circ \mathcal{F}_2 \circ \mathcal{C}_2^p \circ \mathcal{P}_1$$

and a transformation

$$\mathcal{T}_{\Sigma,\Delta} \equiv \mathcal{F}_2 \circ \mathcal{C}_2^p \circ \mathcal{P}_1$$

which maps lists that represent disjoint sums to lists that represent sums.

3.6.3 $\mathcal{T}_{\kappa,\mu}$: MOMENTS $\rightarrow$ CUMULANTS

The implementation of (3.4) again proceeds by example:

$$\Pi_1 \circ \mu_2 \circ \Pi_2 \circ \{\{X, Y\}, \{Z\}\}$$

$$= E[XY]E[Z]$$

$$= \kappa_{X,Y}\kappa_Z + \kappa_X\kappa_Y\kappa_Z$$

$$= \Pi_1 \circ \kappa_2 \circ (\{\{X, Y\}, \{Z\}\} + \{\{X\}, \{Y\}, \{Z\}\})$$

$$= \Pi_1 \circ \kappa_2 \circ \mathcal{F}_1 \circ (\{\{\{X, Y\}\}, \{\{Z\}\}\} + \{\{\{X\}, \{Y\}\}, \{\{Z\}\}\})$$

$$= \Pi_1 \circ \kappa_2 \circ \mathcal{F}_1 \circ \mathcal{P}_2 \circ \{\{X, Y\}, \{Z\}\},$$

from which the identity and transformation can be obtained

$$\Pi_1 \circ \mu_2 \circ \Pi_2 \equiv \Pi_1 \circ \kappa_2 \circ \mathcal{F}_1 \circ \mathcal{P}_2,$$

$$\mathcal{T}_{\kappa,\mu} \equiv \mathcal{F}_1 \circ \mathcal{P}_2,$$

which maps lists that represent moments to lists that represent cumulants.

3.6.4 $\mathcal{T}_{\mu,\kappa}$: CUMULANTS $\rightarrow$ MOMENTS

Similarly, the implementation of (3.4) proceeds as

$$\Pi_1 \circ \kappa_2 \circ \{\{X, Y\}, \{Z\}\}$$

$$= \kappa_{X,Y}\kappa_Z$$

$$= E[XY]E[Z] - E[X]E[Y]E[Z]$$

$$= \Pi_1 \circ \mu_2 \circ \Pi_2 \circ (\{\{X, Y\}, \{Z\}\} - \{\{X\}, \{Y\}, \{Z\}\})$$

$$= \Pi_1 \circ \mu_2 \circ \Pi_2 \circ \mathcal{F}_1 \circ (\{\{\{X, Y\}\}, \{\{Z\}\}\} - \{\{\{X\}, \{Y\}\}, \{\{Z\}\}\})$$

$$= \Pi_1 \circ \mu_2 \circ \Pi_2 \circ \mathcal{F}_1 \circ \mathcal{C}_2^m \circ (\{\{\{X, Y\}\}, \{\{Z\}\}\} + \{\{\{X\}, \{Y\}\}, \{\{Z\}\}\})$$

$$= \Pi_1 \circ \mu_2 \circ \Pi_2 \circ \mathcal{F}_1 \circ \mathcal{C}_2^m \circ \mathcal{P}_2 \circ \{\{X, Y\}, \{Z\}\},$$

from which the identity and transformation can be obtained

$$\Pi_1 \circ \kappa_2 \equiv \Pi_1 \circ \mu_2 \circ \Pi_2 \circ \mathcal{F}_1 \circ \mathcal{C}_2^m \circ \mathcal{P}_2,$$

$$\mathcal{T}_{\mu,\kappa} \equiv \mathcal{F}_1 \circ \mathcal{C}_2^m \circ \mathcal{P}_2.$$

Here $\mathcal{C}_i^m = ((-1)^{l_i \circ \omega - 1} (l_i \circ \omega - 1)!)\omega$ gives another definition for the coefficient operator.

3.6.5 $\mathcal{T}_{\Delta[\mu],\Delta}$: THE EXPECTATION OF AN INDEPENDENT DISJOINT SUM

The expected value of a disjoint sum with independent components is a disjoint sum of expected values:

$$\mu_1 \circ \Delta_1 \circ \Pi_2 \circ \{\{X, Y\}, \{Z\}\} = E\left[\sum_{i \neq j} X_i Y_i Z_j\right]$$

$$= \sum_{i \neq j} E[X_i Y_i] E[Z_j]$$

$$= \Delta_1 \circ \mu_2 \circ \Pi_2 \circ \{\{X, Y\}, \{Z\}\},$$

yielding the identity

$$\mu_i \circ \Delta_i \circ \Pi_2 \equiv \Delta_i \circ \mu_{i+1} \circ \Pi_2.$$

In this case, the transformation $\mathcal{T}_{\Delta[\mu],\Delta}$ is just the identity.

3.6.6 $\mathcal{T}_{\mu,\Delta[\mu]}$: THE EXPECTATION OF AN IID DISJOINT SUM

The expected value of a disjoint sum with IID components collapses to a single term:

$$\Delta_1 \circ \mu_2 \circ \Pi_2 \circ \{\{X, Y\}, \{Z\}\} = \sum_{i \neq j} E[X_i Y_i] E[Z_j]$$

$$= n(n-1) E[XY] E[Z]$$

$$= \mu_1 \circ \Pi_2 \circ (n(n-1)\{\{X, Y\}, \{Z\}\})$$

$$= \mu_1 \circ \Pi_2 \circ \mathcal{C}_1^I \circ \{\{X, Y\}, \{Z\}\},$$

yielding the identity

$$\Delta_1 \circ \mu_2 \circ \Pi_2 \equiv \mu_1 \circ \Pi_2 \circ \mathcal{C}_1^I$$

and the transformation

$$\mathcal{T}_{\mu,\Delta[\mu]} \equiv \mathcal{C}_1^I,$$

where $\mathcal{C}_i^I \circ \omega = c\omega = n!/(n - l_i \circ \omega)!\omega$.

The function l_i is defined in Section 3.6.2. The inverse transformation is simply

$$\mathcal{T}_{\mu,\Delta[\mu]} \equiv \mathcal{C}_1^w,$$

where

$$\mathcal{C}_i^w \circ \omega = (1/c)\omega.$$

In the above definitions, there has been a convenient slight abuse of notation. The transformation denoted by $\mathcal{T}_{\Delta[\mu],\Delta}$ represents the expectation of Δ; the transformation denoted by $\mathcal{T}_{\Delta,\Delta[\mu]}$ represents the unbiased estimation of $\Delta[\mu]$.

3.6.7 $\mathcal{T}_{\mu,\Sigma}$: AN EXAMPLE OF COMPOSING TRANSFORMATIONS

The transformations derived thus far may be used to compute the expectation of products of IID sums. The corresponding transformation is constructed by composing basic transformations to take us along the path $\Sigma \to \Delta \to \Delta[\mu] \to \mu$. Thus, we define

$$\mathcal{T}_{\mu,\Sigma} = \mathcal{T}_{\mu,\Delta[\mu]} \circ \mathcal{T}_{\Delta[\mu],\Delta} \circ \mathcal{T}_{\Delta,\Sigma}.$$

An example of its use is the calculation of the expectation of a product of sums:

$$E\left[\sum XY \sum Z\right] = \mu_1 \circ \Sigma_2 \circ \Pi_2 \circ \{\{X, Y\}, \{Z\}\}$$

$$= \Pi_1 \circ \mu_2 \circ \Pi_2 \circ \mathcal{T}_{\mu,\Sigma} \circ \{\{X, Y\}, \{Z\}\}$$

$$= n(n-1)E[XY]E[Z] + nE[XYZ].$$

3.7   An extended data structure

Other transformations may be composed as well. For example, the fundamental identity of generalized cumulants allows products of joint cumulants of products to be written in terms of products of ordinary cumulants. The cumulant product $\kappa_{WX,Y}\kappa_Z$ can be written in terms of the cumulants of X, Y, Z, and W:

$$\kappa_{WX,Y}\kappa_Z = (E[WXY] - E[WX]E[Y])E[Z]$$

$$= \kappa_{W,Y}\kappa_X\kappa_Z + \kappa_{X,Y}\kappa_W\kappa_Z + \kappa_{W,X,Y}\kappa_Z.$$

The right-hand side is obtained by writing each cumulant in terms of moments and the resulting moments in terms of ordinary cumulants. This is a process that should be automated by composing $\mathcal{T}_{\kappa,\mu}$ and $\mathcal{T}_{\mu,\kappa}$. However, this transformation cannot be used since $\kappa_{WX,Y}\kappa_Z$ cannot be represented by the current data structure used in Section 3.5. It has a more complex structure:

$$\Pi(\kappa[\Pi(\cdots)]),$$

where neither Π nor κ have a single argument. Replacing each of these with the 'list' operator gives

$$\kappa_{WX,Y}\kappa_Z = \Pi_1 \circ \kappa_2 \circ \Pi_3 \circ \omega,$$

where $\omega = \{\{\{W, X\}, \{Y\}\}, \{\{Z\}\}\}$.

Table 3.4. Standard types and functions

Type	Function f_{Type}	$f_{\text{Type}} \circ \omega$
A	$\Pi_1 \circ A_2 \circ \Pi_2 \circ \Pi_3$	$\overline{WXY}\,\overline{Z}$
Δ	$\Delta_1 \circ \Pi_2 \circ \Pi_3$	$W\widetilde{XY}, Z$
μ	$\Pi_1 \circ \mu_2 \circ \Pi_2 \circ \Pi_3$	$E[WXY]E[Z]$
κ	$\Pi_1 \circ \kappa_2 \circ \Pi_3$	$\kappa_{WX,Y}\kappa_Z$
$A[\mu]$	$\Pi_1 \circ A_2 \circ \Pi_2 \circ \mu_3 \circ \Pi_3$	$\overline{E[WX]E[Y]}\,\overline{E[Z]}$
$\Delta[\mu]$	$\Delta_1 \circ \mu_2 \circ \Pi_2 \circ \Pi_3$	$E[W\widetilde{XY}], E[Z]$
$A[\kappa]$	$\Pi_1 \circ A_2 \circ \kappa_2 \circ \Pi_3$	$\overline{\kappa_{WX,Y}}\,\overline{\kappa_Z}$
$\Delta[\kappa]$	$\Delta_1 \circ \kappa_2 \circ \Pi_3$	$\kappa_{W\widetilde{X,Y}}, \kappa_Z$

Similarly, computing expectations for products of sums in the independent but identically distributed case will lead to objects like

$$\sum E[WX]E[Y] \sum E[Z],$$

a product of sums of products of expectations of products of random variables. Since sums and expectations have only one argument, the list ω will accommodate the structure of this object as well; it is nested three times, once for each product. Objects with a more complicated structure rarely arise in statistics (although they do occur as shown in Chapter 9) and hence we adopt ω as our common data structure.

In addition, statistical models are invariably parameterized to some extent by expectations. Sample versions of expectations are averages and hence averages (type A) are adopted as a more natural object than sums. Given our new data structure and the use of averages instead of sums, types are modified in Table 3.4 from those in Table 3.1.

The notation $\widetilde{\cdot}$ denotes a weighted disjoint sum or average; for example,

$$\widetilde{XY, Z} = \frac{1}{n^2} \sum_{i \neq j} X_i Y_i Z_j.$$

3.8 Combining transformations: a vector space

As in the final example of Section 3.6, we continue to compose transformations to permit more complex calculations. The common data structure, exemplified by ω, has changed and so have the transformations. Rather than repeating their derivation by hand, we automate the process in the next section.

The application of any transformation to a list gives a linear combination of lists that have the same structure. In other words, the span of transformations yields a vector space. We need not confine ourselves to the transformations defined thus far. However, there is some merit in identifying a set of fundamental transformations from which all others are derived and confining ourselves to that. The vector space

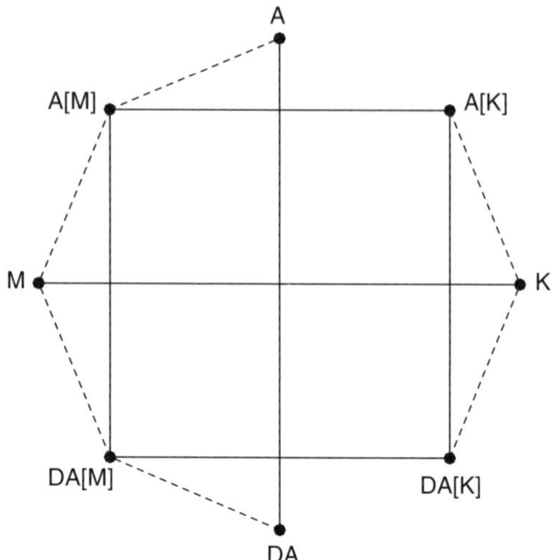

Fig. 3.1.   The span of the basic transformations of types of objects. Transformations appropriate for independent variables are denoted by solid lines; those appropriate for IID variables are denoted by dashed lines.

spanned by this set of transformations is summarized in Figure 3.1. Most of the calculations we collectively call 'mathematical statistics' are constrained to this space. They simply involve moving from vertex to vertex through the composition of transformations. Examples of this follow.

From the connectivity of the figure, it is clear that an object of any type may be transformed into any other type by a sequence of transformations. The transformations between standard types, once defined, may be combined to yield useful results. The sequence of transformations from $A \to B \to C$ is denoted $\mathcal{T}_{C,B,A}$:

$$\mathcal{T}_{C,B,A} \circ \omega = \mathcal{T}_{C,B} \circ \mathcal{T}_{B,A} \circ \omega.$$

The required sequence may be found by following a path in Figure 3.1.

3.8.1   THE FUNDAMENTAL IDENTITY OF CUMULANTS

The more complex data structure allows $\mathcal{T}_{\kappa,\mu}$ and $\mathcal{T}_{\mu,\kappa}$ to be composed to automate the transformation of products of generalized cumulants to ordinary cumulants

$$\mathcal{T}_{\kappa,\kappa} = \mathcal{T}_{\kappa,\mu} \circ \mathcal{T}_{\mu,\kappa}.$$

For example,

$$\Pi_1 \circ \kappa_2 \circ \Pi_3 \circ \{\{\{W, X\}, \{Y\}\}, \{\{Z\}\}\}$$

$$= \kappa_{WX,Y}\kappa_Z$$

$$= \kappa_{W,Y}\kappa_X\kappa_Z + \kappa_{X,Y}\kappa_W\kappa_Z + \kappa_{W,X,Y}\kappa_Z$$

$$= \Pi_1 \circ \kappa_2 \circ \Pi_3 \circ \mathcal{T}_{\kappa,\kappa} \circ \{\{\{W, X\}, \{Y\}\}, \{\{Z\}\}\}.$$

If the initial object is a product of ordinary cumulants then $\mathcal{T}_{\kappa,\kappa}$ acts like an identity function. For example,

$$\Pi_1 \circ \kappa_2 \circ \Pi_3 \circ \{\{\{X\}, \{Y\}\}, \{\{Z\}\}\} = \kappa_{X,Y}\kappa_Z$$
$$= \Pi_1 \circ \kappa_2 \circ \Pi_3 \circ \mathcal{T}_{\kappa,\kappa} \circ \{\{\{X\}, \{Y\}\}, \{\{Z\}\}\}$$
$$= \kappa_{X,Y}\kappa_Z.$$

Writing generalized cumulants in terms of ordinary cumulants is equivalent to the calculation of complementary set partitions as discussed in McCullagh (1987) and Brillinger (1975). The development here permits the calculation of the desired results directly without the introduction of such partitions.

3.8.2 GENERALIZED k-STATISTICS (IID CASE)

Assuming IID variables, composing the transformations $\mathcal{T}_{\Delta,\Delta[\mu]} \circ \mathcal{T}_{\Delta[\mu],\mu}$ automates the derivation of an unbiased estimator of a product of moments. The further application of $\mathcal{T}_{A,\Delta}$ will give the result in terms of averages and preceding $\mathcal{T}_{\Delta[\mu],\mu}$ by $\mathcal{T}_{\mu,\kappa}$ gives an operator for deriving unbiased estimators of generalized cumulants (generalized k-statistics). Defining

$$\mathcal{T}_{A,\kappa} = \mathcal{T}_{A,\Delta} \circ \mathcal{T}_{\Delta,\Delta[\mu]} \circ \mathcal{T}_{\Delta[\mu],\mu} \circ \mathcal{T}_{\mu,\kappa},$$

we compute the unbiased estimates of the second and third cumulants of X:

$$f_A \circ \mathcal{T}_{A,\kappa} \circ \{\{\{X\}, \{X\}\}\} = \frac{\sum X^2}{n-1} - \frac{\left(\sum X\right)^2}{n(n-1)},$$

$$f_A \circ \mathcal{T}_{A,\kappa} \circ \{\{\{X\}, \{X\}, \{X\}\}\}$$
$$= \frac{n\sum X^3}{(n-1)(n-2)} - \frac{3\left(\sum X\right)\left(\sum X^2\right)}{(n-1)(n-2)} + \frac{2\left(\sum X\right)^3}{n(n-1)(n-2)},$$

$$f_A \circ \mathcal{T}_{A,\kappa} \circ \{\{\{X\}, \{X\}, \{X\}, \{X\}\}\}$$
$$= \frac{(n+1)\sum X^4}{(n-1)(n-2)(n-3)} - \frac{4(n+1)\sum X^3 \sum X}{(n-1)(n-2)(n-3)}$$
$$+ \frac{3\left(\sum X^2\right)^2}{(n-2)(n-3)} - \frac{6\left(\sum X\right)^4}{n(n-1)(n-2)(n-3)}.$$

This example indicates the powerful nature of the transformations defined to operate on lists. They provide a simple alternative the methods of Fisher (1928). In the next section we will show how the transformations may be derived from basic identities and develop a small, canonical set of transformations and extend the definitions to more complex objects to handle variables which are not identically distributed and other cases.

3.9 Deriving transformations from identities

In the examples of Section 3.6, we constructed identities involving the basic functions and transformations of our algebra. These identities were in turn used to 'derive' transformations. Details were in the context of a simpler common data structure that has subsequently been complicated to accommodate more complex objects. Rather than repeating these examples, and this process, to identify the transformations of Figure 3.1, the process is automated.

Consider the following example where the transformation $\mathcal{T}_{\kappa,\mu}$ is derived. The distinction between this example and those of Section 3.6 is that the derivation only involves the basic functions and transformations and not lists.

The identities used in this case are

$$1 \ \ \Pi_i \circ \Pi_{i+1} \equiv \Pi_i \circ \mathcal{F}_i,$$

$$2 \ \ \Pi_i \circ \Pi_{i+1} \equiv \Pi_i \circ \Pi_{i+1} \circ \mathcal{D}_i,$$

$$3 \ \ \mu_i \circ \Pi_i \equiv \Pi_i \circ \kappa_{i+1} \circ \mathcal{P}_i,$$

$$4 \ \ \mathcal{P}_i \circ f_j \equiv f_{j+1} \circ \mathcal{P}_i, \ \ i < j,$$

$$5 \ \ \mathcal{F}_i \circ f_j \equiv f_{j-1} \circ \mathcal{F}_i, \ \ i < j.$$

Beginning with f_μ, we apply the above identities until an expression involving f_κ is obtained:

$$f_\mu = \Pi_1 \circ \mu_2 \circ \Pi_2 \circ \Pi_3$$

$$\overset{2}{\to} \Pi_1 \circ \mu_2 \circ \Pi_2 \circ \Pi_3 \circ \mathcal{D}_2$$

$$\overset{3}{\to} \Pi_1 \circ \Pi_2 \circ \kappa_3 \circ \mathcal{P}_2 \circ \Pi_3 \circ \mathcal{D}_2$$

$$\overset{1}{\to} \Pi_1 \circ \mathcal{F}_1 \circ \kappa_3 \circ \mathcal{P}_2 \circ \Pi_3 \circ \mathcal{D}_2$$

$$\overset{5}{\to} \Pi_1 \circ \kappa_2 \circ \mathcal{F}_1 \circ \mathcal{P}_2 \circ \Pi_3 \circ \mathcal{D}_2$$

$$\overset{4}{\to} \Pi_1 \circ \kappa_2 \circ \mathcal{F}_1 \circ \Pi_4 \circ \mathcal{P}_2 \circ \mathcal{D}_2$$

$$\overset{5}{\to} \Pi_1 \circ \kappa_2 \circ \Pi_3 \circ \mathcal{F}_1 \circ \mathcal{P}_2 \circ \mathcal{D}_2$$

$$= f_\kappa \circ \mathcal{F}_1 \circ \mathcal{P}_2 \circ \mathcal{D}_2$$

$$= f_\kappa \circ \mathcal{T}_{\kappa,\mu},$$

where $\overset{i}{\to}$ denotes the computed result of the application of identity i. Removing the functions f_μ and f_κ from each side leaves us with the transformation

$$\mathcal{T}_{\kappa,\mu} \equiv \mathcal{F}_1 \circ \mathcal{P}_2 \circ \mathcal{D}_2.$$

Due to the more complex data structure, more than one identity was required to derive the transformation $\mathcal{T}_{\kappa,\mu}$. It is this process of combining identities to yield transformations that is automated. The exact identities required in the derivation and the order in which they should be applied must be identified by the user. Since transformations are derived from identities automatically, we need only get the identities right.

In general, identities are applied to the function form of one type, say f_A, until we arrive at an expression that involves the functional form of a second type, say f_B, that is, until we get a statement of equivalence of the form

$$f_1 \circ \cdots \circ f_j \equiv f_{j+1} \circ \cdots \circ f_l \circ \mathcal{T}_k \circ \cdots \circ \mathcal{T}_l$$

or rather

$$f_A \equiv f_B \circ \mathcal{T}_{B,A},$$

where $f_A = f_1 \circ \cdots \circ f_j$, $f_B = f_{j+1} \circ \cdots \circ f_l$ and $\mathcal{T}_{B,A} = \mathcal{T}_k \circ \cdots \circ \mathcal{T}_l$. Removing the functions f_A and f_B from each side leaves us with $\mathcal{T}_{B,A}$. This process is automated by a utility and examples of its use are given in Section 3.11. However, preceding this all elementary identities must be derived.

3.10 Identities

Functions and transformations are related by a small number of elementary identities. These identities will be used to derive all fundamental transformations automatically. General definitions of all of these identities, in terms of the basic functions and transformations, are presented in Table 3.5. For exposition, the identities of Section 3.9 are derived here with the aid of examples. We proceed in a manner similar to the examples of Section 3.6.

3.10.1 MULTIPLICATION IS ASSOCIATIVE

For our purposes, the associative law of multiplication is encapsulated by two identities,

$$(X)(YZ) = (XYZ),$$

$$(X)(YZ) = (X)(Y)(Z).$$

In terms of our operators, an example of the first is

$$\Pi_1 \circ \Pi_2 \circ \{\{X\}, \{Y, Z\}\} = ((X)(YZ)) = XYZ$$

$$= \Pi_1 \circ \{X, Y, X\}$$

$$= \Pi_1 \circ \mathcal{F}_1 \circ \{\{X\}, \{Y, Z\}\}.$$

Now if we examine the two extremes of this expression and drop the list $\{\{X\}, \{Y, Z\}\}$ in each case, we arrive at an equivalent identity expressed in terms of our operators, namely $\Pi_1 \circ \Pi_2 \equiv \Pi_1 \circ \mathcal{F}_1$. The general identity can then be inferred to be

$$\Pi_i \circ \Pi_{i+1} \equiv \Pi_i \circ \mathcal{F}_i.$$

It is in this manner that identities are rewritten in terms of our operators. An example of the second identity is

$$\Pi_1 \circ \Pi_2 \circ \{\{X\}, \{Y, Z\}\} = ((X)(YZ)) = (X)(Y)(Z)$$
$$= \Pi_1 \circ \Pi_2 \circ \{\{X\}, \{Y\}, \{Z\}\}$$
$$= \Pi_1 \circ \Pi_2 \circ \mathcal{D}_1 \circ \{\{X\}, \{Y, Z\}\},$$

yielding the general identity

$$\Pi_i \circ \Pi_{i+1} \equiv \Pi_i \circ \Pi_{i+1} \circ \mathcal{D}_i.$$

3.10.2 MOMENTS TO CUMULANTS

Similarly, the identity (3.4) gives moments in terms of cumulants. The example

$$\mu_{XY} = \kappa_{X,Y} + \kappa_X \kappa_Y$$

may be rewritten in terms of our operators as

$$\mu_1 \circ \Pi_1 \circ \{X, Y\} = \mu_{XY} = \kappa_{X,Y} + \kappa_X \kappa_Y$$
$$= \Pi_1 \circ \kappa_2 \circ (\{\{X, Y\}\} + \{\{X\}, \{Y\}\})$$
$$= \Pi_1 \circ \kappa_2 \circ \mathcal{P}_1 \circ \{X, Y\},$$

yielding

$$\mu_i \circ \Pi_i \equiv \Pi_i \circ \kappa_{i+1} \circ \mathcal{P}_i.$$

3.10.3 IDENTITIES DERIVED FROM THE ALGEBRA

To derive a fundamental transformation from a set of identities, we must write an expression as a sequence of functions followed by a sequence of transformations. Doing so requires the ability to commute functions and transformations. For example, the statement in the above example

$$f_\mu = \Pi_1 \circ \mathcal{F}_1 \circ \kappa_3 \circ \mathcal{P}_2 \circ \Pi_3 \circ \mathcal{D}_2$$

does not allow the identification of $\mathcal{T}_{\kappa,\mu}$. To get $\mathcal{T}_{\kappa,\mu}$, we must commute κ with $\mathcal{F}$ and Π with both $\mathcal{F}$ and $\mathcal{P}$. So, we need to study some of the algebraic properties of our own operators.

When $i < j$ some functions and transformations commute, but one must account for the effect of the transformation. The partition operator adds a level to a list and so we have the identity

$$\mathcal{P}_i \circ f_j \equiv f_{j+1} \circ \mathcal{P}_i.$$

For example,

$$\mathcal{P}_1 \circ \mu_2 \circ \{X, Y\} = \mathcal{P}_1 \circ \{E[X], E[Y]\}$$
$$= \{\{E[X], E[Y]\}\} + \{\{E[X]\}, \{E[Y]\}\}$$
$$= \mu_3 \circ \{\{X, Y\}\} + \{\{X\}, \{Y\}\}$$
$$= \mu_3 \circ \mathcal{P}_1 \circ \{X, Y\}.$$

Similarly, $\mathcal{F}$ reduces a list by a level and so

$$\mathcal{F}_i \circ f_j \equiv f_{j-1} \circ \mathcal{F}_i.$$

For example,

$$\mathcal{F}_1 \circ \mu_3 \circ \{\{X\}, \{Y\}\} = \mathcal{F}_1 \circ \{\{E[X]\}, \{E[Y]\}\}$$
$$= \{E[X], E[Y]\}$$
$$= \mu_2 \circ \{X, Y\}$$
$$= \mu_2 \circ \mathcal{F}_1 \circ \{\{X\}, \{Y\}\}.$$

3.10.4 ALL ELEMENTARY IDENTITIES

All of the elementary identities, numbered for future reference, are summarized in Table 3.5. Justification of each identity in the above manner is left as an exercise.

Table 3.5. Identities involving functions and transformations

Number	Identity
1	$\Pi_i \circ \Pi_{i+1} \equiv \Pi_i \circ \mathcal{F}_i$
2	$\Pi_i \circ \Pi_{i+1} \equiv \Pi_i \circ \Pi_{i+1} \circ \mathcal{D}_i$
6	$A_i \circ \mu_i = \mu_i \circ A_i$
3	$\mu_i \circ \Pi_i \equiv \Pi_i \circ \kappa_{i+1} \circ \mathcal{P}_i$
7	$\kappa_i \equiv \Pi_i \circ \mu_{i+1} \circ \Pi_{i+1} \circ \mathcal{C}^{P(1)}{}_i \circ \mathcal{P}_i$
8	$\Pi_i \circ A_{i+1} \equiv \Delta_i \circ \Pi_{i+1} \circ \mathcal{C}^{V(1)}{}_i \circ \mathcal{P}_i \circ \mathcal{C}^{A(1)}{}_i$
9	$\Delta_i \equiv \Pi_i \circ A_{i+1} \circ \Pi_{i+1} \circ \mathcal{C}^{V(1)}{}_i \circ \mathcal{C}^{P(1)}{}_i \circ \mathcal{P}_i \circ \mathcal{C}^{A(1)}{}_i$
	for commuting functions and transformations $i < j$:
16	$\mathcal{C}^{\cdot(k)}{}_i \circ f_j \equiv f_j \circ \mathcal{C}^{\cdot(k)}_i$
4	$\mathcal{P}_i \circ f_j \equiv f_{j+1} \circ \mathcal{P}_i$
5	$\mathcal{F}_i \circ f_j \equiv f_{j-1} \circ \mathcal{F}_i$
	for variables with independent components:
10	$\mu_i \circ \Delta_i \equiv \Delta_i \circ \mu_{i+1}$
11	$\kappa_i \circ A_i \equiv A_i \circ \mathcal{C}^{V(1)}{}_i \circ \kappa_i \circ \mathcal{C}^{A(1)}{}_{i+1}$
12	$\Delta_i \circ \mu_{i+1} \equiv \mu_i \circ \Delta_i$
	for variables with IID components:
13	$\Pi_i \circ \mu_{i+1} \equiv \Pi_i \circ A_{i+1} \circ \mu_{i+1}$
14	$\Delta_i \circ \mu_{i+1} \equiv \Pi_i \circ \mu_{i+1} \circ \mathcal{C}^{M(1)}{}_i$
15	$\Pi_i \circ \mu_{i+1} \equiv \Delta_i \circ \mu_{i+1} \circ \mathcal{C}^{W(1)}{}_i$

3.11   Computer algebra proofs

The identities of the previous section can be defined in a computer algebra environment as replacement statements. An operator can be defined to derive transformations by using identities to change the form of a function for one type into an expression involving a function for another type. Such an operator would automate a mathematical proof.

In the three examples that follow, transformations are derived from identities. Although the process can be automated in the above manner, providing a detailed derivation similar to that of Section 3.9 helps the reader to understand the operators for computer algebra proofs.

3.11.1   CUMULANTS OF PRODUCTS OF AVERAGES

Many estimates may be expressed in terms of products of averages: type A. The variance is a simple example. To calculate cumulants of averages, we first define the type $\kappa[A]$ representing a cumulant with components which are products of averages: $\kappa[A] = \kappa_1 \circ \Pi_2 \circ A_3 \circ \Pi_3$. Thus, the list $\{\{\{X\}, \{Y\}\}, \{\{Z\}\}\}$ represents the cumulant $\kappa_{\overline{X}\ \overline{Y},\overline{Z}}$.

By applying identities one at a time, it is found that the basic identities 7, 4, 1, 5, 8, 4, 10, 14, 1, 5 applied to type $\kappa[A]$ produce type μ, thereby yielding a transformation to express the cumulant in terms of moments:

$$f_{\kappa[A]} = \kappa_1 \circ \Pi_2 \circ A_3 \circ \Pi_3$$

$$\overset{7}{\to} \Pi_1 \circ \mu_2 \circ \Pi_2 \circ \mathcal{C}^P{}_1 \circ \mathcal{P}_1 \circ \Pi_2 \circ A_3 \circ \Pi_3$$

$$\overset{4}{\to} \Pi_1 \circ \mu_2 \circ \Pi_2 \circ \Pi_3 \circ A_4 \circ \Pi_4 \circ \mathcal{C}^P{}_1 \circ \mathcal{P}_1$$

$$\overset{1}{\to} \Pi_1 \circ \mu_2 \circ \Pi_2 \circ \mathcal{F}_2 \circ A_4 \circ \Pi_4 \circ \mathcal{C}^P{}_1 \circ \mathcal{P}_1$$

$$\overset{5}{\to} \Pi_1 \circ \mu_2 \circ \Pi_2 \circ A_3 \circ \Pi_3 \circ \mathcal{F}_2 \circ \mathcal{C}^P{}_1 \circ \mathcal{P}_1$$

$$\overset{8}{\to} \Pi_1 \circ \mu_2 \circ \Delta_2 \circ \Pi_3 \circ \mathcal{C}^V{}_2 \circ \mathcal{P}_2 \circ \mathcal{C}^A{}_2 \circ \Pi_3 \circ \mathcal{F}_2 \circ \mathcal{C}^P{}_1 \circ \mathcal{P}_1$$

$$\overset{4}{\to} \Pi_1 \circ \mu_2 \circ \Delta_2 \circ \Pi_3 \circ \Pi_4 \circ \mathcal{C}^V{}_2 \circ \mathcal{P}_2 \circ \mathcal{C}^A{}_2 \circ \mathcal{F}_2 \circ \mathcal{C}^P{}_1 \circ \mathcal{P}_1$$

$$\overset{10}{\to} \Pi_1 \circ \Delta_2 \circ \mu_3 \circ \Pi_3 \circ \Pi_4 \circ \mathcal{C}^V{}_2 \circ \mathcal{P}_2 \circ \mathcal{C}^A{}_2 \circ \mathcal{F}_2 \circ \mathcal{C}^P{}_1 \circ \mathcal{P}_1$$

$$\overset{14}{\to} \Pi_1 \circ \Pi_2 \circ \mu_3 \circ \mathcal{C}^M{}_2 \circ \Pi_3 \circ \Pi_4 \circ \mathcal{C}^V{}_2 \circ \mathcal{P}_2 \circ \mathcal{C}^A{}_2 \circ \mathcal{F}_2 \circ \mathcal{C}^P{}_1 \circ \mathcal{P}_1$$

$$\overset{1}{\to} \Pi_1 \circ \mathcal{F}_1 \circ \mu_3 \circ \mathcal{C}^M{}_2 \circ \Pi_3 \circ \Pi_4 \circ \mathcal{C}^V{}_2 \circ \mathcal{P}_2 \circ \mathcal{C}^A{}_2 \circ \mathcal{F}_2 \circ \mathcal{C}^P{}_1 \circ \mathcal{P}_1$$

$$\overset{5}{\to} \Pi_1 \circ \mu_2 \circ \Pi_2 \circ \Pi_3 \circ \mathcal{F}_1 \circ \mathcal{C}^M{}_2 \circ \mathcal{C}^V{}_2 \circ \mathcal{P}_2 \circ \mathcal{C}^A{}_2 \circ \mathcal{F}_2 \circ \mathcal{C}^P{}_1 \circ \mathcal{P}_1$$

$$= f_\mu \circ \mathcal{F}_1 \circ \mathcal{C}^M{}_2 \circ \mathcal{C}^V{}_2 \circ \mathcal{P}_2 \circ \mathcal{C}^A{}_2 \circ \mathcal{F}_2 \circ \mathcal{C}^P{}_1 \circ \mathcal{P}_1$$

$$= f_\mu \mathcal{T}_{\mu,\kappa[A]},$$

where each $C^{\bullet(1)}$ has been abbreviated to $C^{\cdot}$. The use of the superscript indicating the level permits the interchange of P and C, so that some constants may be collected. Thus, the implicitly defined transformation is

$$\mathcal{T}_{\mu,\kappa[A]} = \mathcal{F}_1 \circ C^{M(1)}{}_2 \circ C^{V(1)}{}_2 \circ C^{A(2)}{}_2 \circ \mathcal{P}_2 \circ \mathcal{F}_2 \circ C^{P(1)}{}_1 \circ \mathcal{P}_1$$

3.11.2 EXAMPLE

The Gaussian maximum likelihood estimate of variance is $\overline{X^2} - \overline{X}^2$, an object of type A, represented by $V = \{\{\{X, X\}\}\} - \{\{\{X\}\}, \{\{X\}\}\}$. Since the innermost level of the structure of type A is not required here, we flatten this level and compute $v = \mathcal{F}_2 \circ V = \{\{X, X\}\} - \{\{X\}, \{X\}\}$ and represent *its* variance as an object of type $\kappa[A]$ by $\{v, v\} = \{\{\{X, X\}\} - \{\{X\}, \{X\}\}, \{\{X, X\}\} - \{\{X\}, \{X\}\}\}$. The linear nature of lists will result in the expansion of this expression. This cumulant can be expressed in terms of simple cumulants by applying $\mathcal{T}_{\kappa,\mu,\kappa[A]}$:

$$\mathcal{T}_{\kappa,\mu,\kappa[A]} \circ \{\{\{X, X\}\} - \{\{X\}, \{X\}\}, \{\{X, X\}\} - \{\{X\}, \{X\}\}\}$$
$$\to 2(n^{-1} - n^{-2})\kappa^2_{X,X} + (n^{-1} - 2n^{-2} + n^{-3})\kappa_{X,X,X,X}.$$

3.11.3 THE FUNDAMENTAL TRANSFORMATIONS

Only some transformations need to be defined, as these basic transformations may be composed to yield the remaining ones. Table 3.6 was produced by computer algebra proof.

The constant operator C, as defined for example in (3.6), has been generalized to a form $C^{\cdot(i)}{}_j$ where the dependence of C is on the total number of components at level i and the operator is applied to level j. This leads to simpler definitions:

$$C^{P(k)} \circ \omega = (-1)^{(l_k-1)}(l_k - 1)!\omega,$$

$$C^{A(k)} \circ \omega = n^{-l_k}\omega,$$

$$C^{V(k)} \circ \omega = n^{l_k}\omega,$$

$$C^{m(k)} \circ \omega = n^{-l_k}\frac{n!}{(n - l_k)!}\omega,$$

$$C^{W(k)} \circ \omega = n^{l_k}\left(\frac{n!}{(n - l_k)!}\right)^{-1}\omega,$$

where l_k denotes the total length of components at level k of ω. In addition, this generalization permits the transposition of operators so that, for example, $\mathcal{P} \circ C$ may be expressed in the form $C \circ \mathcal{P}$. Such re-expressions are useful for some simplification of operators. It is left as an exercise for the reader to verify each transformation.

Table 3.6. Derivation of transformations

Type	Identities used	Transformation
$\mathcal{T}_{\Delta,A}$	1, 8	$\mathcal{C}^{V(1)}{}_1 \circ \mathcal{C}^{A(2)}{}_1 \circ \mathcal{F}_3 \circ \mathcal{P}_1$
$\mathcal{T}_{A,\Delta}$	1, 9, 2	$\mathcal{D}_2 \circ \mathcal{C}^{V(1)}{}_1 \circ \mathcal{C}^{P(1)}{}_2 \circ \mathcal{C}^{A(2)}{}_1 \circ \mathcal{F}_3 \circ \mathcal{P}_1$
$\mathcal{T}_{\Delta[\mu],\Delta}$	10	$\mathcal{I}$
$\mathcal{T}_{\Delta,\Delta[\mu]}$	12	$\mathcal{I}$
$\mathcal{T}_{\mu,\Delta[\mu]}$	14	$\mathcal{C}^{m(1)}{}_1$
$\mathcal{T}_{\Delta[\mu],\mu}$	15	$\mathcal{C}^{W(1)}{}_1$
$\mathcal{T}_{A[\mu],\Delta[\mu]}$	1, 9	$\mathcal{C}^{V(1)}{}_1 \circ \mathcal{C}^{P(1)}{}_2 \circ \mathcal{C}^{A(2)}{}_1 \circ \mathcal{F}_1 \circ \mathcal{P}_1$
$\mathcal{T}_{\kappa,\mu}$	2, 3, 1	$\mathcal{F}_1 \circ \mathcal{P}_2 \circ \mathcal{D}_2$
$\mathcal{T}_{\mu,\kappa}$	7, 1	$\mathcal{F}_1 \circ \mathcal{C}^{m(1)}{}_2 \circ \mathcal{P}_2$

3.11.4 CUMULANTS OF GENERALIZED k-STATISTICS

The methodology described here requires no distinction between standard and generalized k-statistics, the latter being unbiased estimates of generalized cumulants which may involve products, as in $\kappa_{XY,Z}$. Indeed, the transformation, $\mathcal{T}_{A,\Delta,\Delta[\mu],\mu,\kappa}$, of the previous section produces unbiased estimates of products of generalized cumulants.

Cumulants of k-statistics may be simply computed by defining a type, $\kappa[k]$, to display these. For example, the unbiased estimate of the covariance $\kappa_{X,Y}$ is denoted by $k_{X,Y}$ and the variance of this k-statistic is denoted by $\kappa_{k_{X,Y},k_{X,Y}}$.

The basic identities are used to compute the required transformations to other forms. However, the calculations are simplified by defining a new, special type. Since the k-statistics are products of averages, type A, cumulants of these will be denoted by type $\kappa[A]$ defined by $\kappa[A] = \kappa_1 \circ \Pi_2 \circ A_3 \circ \Pi_3$. Application of the identities used to define $\mathcal{T}_{\mu,\kappa}$ and $\mathcal{T}_{A,\Delta,\Delta[\mu],\mu}$ to type $\kappa[A]$ yield type μ, thereby defining the transformation $\mathcal{T}_{\mu,\kappa[A]}$. Thus, the transformation $\mathcal{T}_{\kappa,\mu,\kappa[A]}$ applied to a list representing the cumulant of k-statistics yields the cumulant expressed in terms of cumulants of the component random variables.

To illustrate the simplicity of this approach, we define

$$k_{X,Y,\dots} = \mathcal{T}_{A,\mu,\Delta,\Delta[\mu],\mu,\kappa} \circ \kappa_{X,Y,\dots}.$$

The operator $\mathcal{T}_{A,\mu,\Delta,\Delta[\mu],\mu,\kappa}$ which operates on a list may be simply modified to operate on the *components* of a list by just increasing the index of each operator by 1. If the modified operator is applied to the nested list representing a cumulant of k-statistics, the result represents a cumulant of products of averages, type $\kappa[A]$. Combining the above with $\mathcal{T}_{\kappa,\mu}$ and $\mathcal{T}_{\mu,\kappa[A]}$ yields a transformation from $\kappa[k]$ to κ denoted by $\mathcal{T}_{\kappa,\kappa[k]}$. To simplify the display of the resulting expressions, we assume that all variables have expectation 0.

For example, the fourth cumulant of $k_{X,X}$, the unbiased estimate of the variance of X is denoted by

$$\mathcal{T}_{\kappa,\kappa[k]}[\{V, V, V, V\}],$$

where $V = \{\{\{X\}, \{X\}\}\}$. This yields

$$\frac{48\,\kappa_{X,X^4}}{(-1+n)^3} + \frac{96\,(-2+n)\,\kappa_{X,X}\,\kappa_{X,X,X^2}}{(-1+n)^3\,n} + \frac{144\,\kappa_{X,X^2}\,\kappa_{X,X,X,X}}{(-1+n)^2\,n}$$

$$+ \frac{8\,(6-9n+4n^2)\,\kappa_{X,X,X,X^2}}{(-1+n)^3\,n^2} + \frac{32\,(-2+n)\,\kappa_{X,X,X}\,\kappa_{X,X,X,X,X}}{(-1+n)^2\,n^2}$$

$$+ \frac{24\,\kappa_{X,X}\,\kappa_{X,X,X,X,X,X}}{(-1+n)\,n^2} + \frac{\kappa_{X,X,X,X,X,X,X,X}}{n^3}.$$

(We have left the rather curious term $(-1+n)$ as it was produced by the computer.)

In the special case where the k-statistics are unbiased estimates of *simple* cumulants, the result may also be obtained by applying pattern functions to a complementary set partition as discussed in McCullagh (1987) and implemented in Yong Wang (1992). The methods presented here are more general and apply to generalized k-statistics and exploit the symmetries that result when variables are repeated in expressions, as in the example above. This avoids the computation associated with the 2145 terms of the complementary set partition of $\{\{r, s\}, \{t, u\}, \{v, w\}, \{x, y\}\}$, to which the pattern function would be applied.

The covariance of the generalized k-statistics $k_{XY,X}$ and k_Y is found simply by applying the transformation $\mathcal{T}_{\kappa,\kappa[k]}$:

$$\mathcal{T}_{\kappa,\kappa[k]}[\{\{\{\{X, Y\}, \{Y\}\}\}, \{\{\{Y\}\}\}\}] = \frac{\kappa_{Y,Y}\,\kappa_{X,X,Y}}{n}.$$

The covariance of the unbiased estimate of a variance and of its square, $\kappa_{k_{X,X},k_{X,X}^2}$ is

$$V = \{\{X\}, \{X\}\},$$

$$\mathcal{T}_{\kappa,\kappa[k]}[\{\{V, V\}, \{V\}\}] = \frac{4\,\kappa_{X,X^3}}{-1+n} - \frac{2\,\kappa_{X,X,X^2}}{(-1+n)\,n} + \frac{2\,\kappa_{X,X}\,\kappa_{X,X,X,X}}{n},$$

where again the term $(-1 + n)$ generated by the computer appears.

3.12 The connection with the rest of the book

The procedures used for the calculation of moments, cumulants, and their unbiased estimates are common to many inference methods. These come in the form of exact and asymptotic calculations. The latter requires operators for 'linearizing' a statistic, the topic of the next chapter. In the remaining chapters, we show how the operators described here can be used to automate otherwise labourious calculations in parametric and non-parametric inference, likelihood, ideal bootstrap, Edgeworth and saddle-point methods, M-estimation, and sample survey theory. With the exception of sample survey theory, where multiple-stage and multiple-phase designs are encountered, no further development of the operators discussed here is required. For survey theory, the extension of the operators is trivial. Thus, the technology of this chapter is henceforth buried in a 'black box'.

3.13   Bibliographical notes

Partitions appear throughout the literature for the computation of moments, cumulants, k-statistics, and so forth. Kendall and Stuart (1977) is one of the most familiar sources. The notion of a full partition used here is closely related to standard ordered partitions (Kendall 1993, McCullagh and Wilks 1988) which is a stricter version of the above as it imposes ordering the blocks such that the first entry of each block has ascending order. Since this condition is not required for the computations in this book it is not imposed.

Complementary set partitions are discussed extensively in McCullagh (1987), including references to their first appearance in the literature in James (1958) and Leonov and Shirayev(1959). The discussion is accompanied by dense jargon that includes Hasse diagrams, Möbius inversion, elementary lattice theory, least upper bounds, and so forth. A similar treatment can be found in Barndorff-Nielsen and Cox (1989). The beauty of such text is in itself useful. It deepens our understanding. However, for the purposes of computations, these concepts can be avoided. Brillinger (1975) uses complementary set partitions for theoretical developments in time series. Smith and Field (2000) use the methods of this chapter for cumulant calculations in a time series problem. The reduction of a complementary set partition to a set of equivalence classes requires the use of an intersection matrix which is discussed in Chapter 10.

Computational techniques using partitions are discussed in at least McCullagh (1987), McCullagh and Wilks (1988), Kendall (1994), Stafford (1994), Andrews and Stafford (1998) and Bellhouse, Philips, and Stafford (1997). The use of identities including those to permute operators is discussed in Andrews (2000). The identities lead to the methods of Andrews and Stafford (1998).

3.14   Problems

1. Using known results, discuss and implement ways of testing the operators of this chapter. For example, $\mathcal{T}_{A,\Delta} \circ \mathcal{T}_{\Delta,A}$ should be the identity.
2. Modify $f_\kappa \circ \mathcal{T}_{\kappa,\mu} \circ \mathcal{T}_{\mu,\kappa}$ so that it explicitly derives a complementary set partition.
3. Use the operators of Chapters 2 and 3 to show that the covariance of a sample mean and sample variance is zero for a random sample from the normal distribution.
4. Design an operator for computing joint cumulants of averages when sampling without replacement from finite population.
5. Use the methods of this chapter to verify Table 4.1 of McCullagh (1987).
6. Bootstrap or plug-in estimates of moments are expressions in which expectations are replaced by averages. These correspond to the identity transformation $\mathcal{T}_{A,A[\mu]}$. Design an operator for computing such bootstrap estimates of cumulants. Use it to compute estimates of the variance of a sample mean and the bias of a sample variance.

7. The expectation of a product of averages may be computed using either the relations between moments and cumulants or the relations between averages and disjoint averages. Show that these involve exactly the same computations.

8. There is more than one way to derive the transformation $\mathcal{T}_{\mu,\kappa}[A]$. Suppose in Section 3.11.1 we replace identity 8 with 3, that is, we write moments of averages in terms of cumulants of averages, and then invoke identity 11, and so forth. What further identities are needed to derive the transformation? Is it the same algorithm as in Section 3.11.1? If not, is it better?

9. Suppose, instead, we wished to derive the transformation $\mathcal{T}_{\kappa,\kappa}[A]$. In the manner of the previous problem, discuss and compare two ways of doing this.

10. In the manner of Section 3.10, verify the identities of Table 3.5.

11. Use the methods of Chapters 2 and 3 to compute cumulants of the least-squares estimates for the slope and intercept in simple linear regression.

12. Under some null hypothesis, use the methods of Chapters 2 and 3 to compute cumulants for the standardized mean in the 'one-sample problem'. Assume the variance, σ^2, is known. Repeat these calculations for the same standardized mean where the null hypothesis is false and again where the specified value of σ^2 is wrong.

13. For any density $f_\theta(X)$, with log-likelihood $l(\theta) = \log\{f_\theta(X)\}$, we have the identity

$$\int e^{l(\theta)}\, dX = 1.$$

Assuming integration and differentiation can commute, further identities can be derived by differentiating both sides with respect to θ. In the manner of Section 3.3 establish that the resulting identities depend on full partitions. Develop an algorithm for deriving these identities.

14. The uses of Laplace–Stieltjes transforms (LSTs) and probability generating functions (PGFs) are quite common in queueing theory. For the standard M/G/1 queue with arrival rate λ, service rate μ, and $\rho = \lambda/\mu < 1$, it is well known that the Laplace–Stieltjes transform is

$$\psi(z) = z\phi\left(\lambda - \lambda\psi(z)\right).$$

Moments for the busy period of this queue can be computed from the derivatives of ψ; however, these derivatives are implicit. Develop an algorithm that uses full partitions to evaluate these derivatives at zero, thus avoiding the expression of the derivative itself.

4
Asymptotic expansions

4.1 Introduction

This chapter describes algorithms for deriving asymptotic expansions. We begin
with a formal definition of an expansion in the manner of Barndorff-Nielsen and
Cox (1989), Chapter 3. Let $\beta_0, \beta_1, \ldots$ be a sequence such that $\beta_0 = 1$ and $\beta_i = o(\beta_{i-1})$. For a quantity g, assume that there exist coefficients $\gamma_0, \gamma_1, \gamma_2, \ldots$ such
that

$$g = \gamma_0 + \gamma_1\beta_1 + \cdots + \gamma_k\beta_k + O(\beta_{k+1})$$

or

$$g = \gamma_0\{1 + \gamma_1/\gamma_0\beta_1 + \cdots + \gamma_k/\gamma_0\beta_k\} + O(\beta_{k+1}).$$

In either case, the right-hand side is called the asymptotic expansion of g with
leading term γ_0.

In statistics, the sequence β_i depends on some measurement of information
about an unknown parameter. The sample size n is an example. Throughout this
chapter, and in the rest of the book, we assume that

$$\beta_0 = 1, \quad \beta_1 = 1/\sqrt{n}, \quad \beta_2 = 1/n, \quad \ldots, \quad \beta_i = (1/\sqrt{n})^i, \quad \ldots$$

In situations where this is not the case, the algorithms of this chapter still apply.
To emphasize this point, we make β_i implicit in our notation by setting $g_i = \gamma_i\beta_i$.
We refer to g_i throughout as the term of order i.

The technical objective of an asymptotic analysis involving expansions is the
determination of the terms $g_0, g_1, \ldots$ The rationales for examining these terms
and their interpretation are varied and of primary importance. While deriving
these terms is a necessary part of the process, it is, in a sense, the easiest part.
Unfortunately, it is often not easy and the algorithms of this chapter are an attempt
to alleviate this difficulty. There is no guarantee that an expansion is valid or that
the results will be useful. We follow the tradition, long established in computing,
and assume that the user knows what s/he wants to compute and wishes to do it
correctly and efficiently.

The computations associated with asymptotic expansions may be broken down
into a small number of basic procedures or rules. These may be implemented
directly and certain computational advantages are realized when this is done for

the expansion as a whole. Alternatively, closer examination of the basic rules leads to defining the procedures in terms of integer partitions in the same way that the procedures of the previous chapter relied on set partitions. Integer partitions provide an elegant means of gaining insight into a problem and improve efficiency because they identify terms uniquely and can be saved.

4.2 Procedures for manipulating expansions

We begin with procedures for calculating expansions to some specific order k: $g = \sum_{i=0}^{k} g_i$. As in Chapter 3, and for computational purposes, expansions are represented by lists where the length of the list gives the order of the expansion; for example, g becomes $\{g_0, \ldots, g_k\}$. The procedures described below operate on such lists to produce new lists representing results of operations on expansions. In addition, procedures may be defined in an object-oriented way, so the order of a result depends only on the order of the arguments. The basic procedures for computing expansions are outlined below.

4.2.1 MULTIPLICATION BY A CONSTANT

If a is a constant, independent of n, and g is an expansion, the k terms of the expansion of $h = ag$ are given by

$$h_i = ag_i,$$

since multiplication by a does not affect the asymptotic order of the terms.

4.2.2 ADDITION

If g and h are expansions to order k, the terms of the expansion of $t = g + h$ are given by

$$t_i = g_i + h_i,$$

because addition of terms of the same order produces a result of that order.

These first two rules are the same as those for the scalar multiplication and addition of lists (vectors) common to most packages.

4.2.3 PRODUCTS

If g and h are expansions to order k, the terms of the expansion of $t = g \times h$ are given by

$$t_i = \sum_{j=0}^{i} g_j \vee h_{j-i}, \tag{4.1}$$

where $\vee$ denotes the operation of multiplication appropriate for the type of g_j and h_{j-i}. For example, $t_2 = g_0 h_2 + g_1 h_1 + g_2 h_0$. It is important for the reader to realize that '$\times$' is a user-defined operator whose inputs are lists (expansions) and whose output is a new list (expansion) with entries defined by the above rule. In keeping with the object-oriented approach mentioned above, the length of the output list depends on the common length of the input lists.

In Chapter 3, types of objects were defined in terms of lists. For all of the objects of Table 3.4, except those involving disjoint averages, the outer list denoted a product. If multiplication is restricted to these types, the multiplication $g_j \vee h_{j-i}$ in (4.1) involves just joining the two lists g_j and h_{j-i}.

4.2.4 SMOOTH FUNCTIONS OF EXPANSIONS

A large class of asymptotic expansions result from series expansions of functions. Taylor's theorem asserts that the behaviour of a function, say $f(g)$, in the neighbourhood of some point g_0, can be well approximated by an expansion that involves the function and its derivatives. Namely,

$$f(g) = \sum_{i=0}^{\infty} f^{(i)}(g_0)\, \delta^i/i!, \tag{4.2}$$

where $f^{(i)}$ denotes the ith derivative of f and $\delta = g - g_0$. We assume that g is close enough to g_0 and the derivatives are sufficiently small that the series converges.

Denoting the expansion of $f^{(r)}(g_0)$ to order k as f^r and similarly for δ we have

$$f^r = \sum_{i=0}^{k} f_i^r \quad \text{and} \quad \delta = \sum_{i=1}^{k} \delta_i = \sum_{i=1}^{k} g_i,$$

since $g_i = \delta_i$ if $i > 0$. The expansion of $f(g)$ truncated to order k is then given by

$$f(g) = \sum_{i=0}^{k} f^i \delta^i/i!, \tag{4.3}$$

which relies wholly on procedures for constant multiplication, addition, and products of expansions. Since δ^i is of order i, terms arising when $i > k$ are of smaller order and do not contribute to the expansion.

An efficient implementation of (4.3) is as follows:

1. initialize: $\{\delta = g;\ \delta_0 = 0;\ t = \delta;\ h = f^0 + f^1 \times \delta\}$,
2. for $i = 2$ to $(k-1)$ do: $\{t = (1/i)\delta \times t;\ h = h + f^i \times t\}$.

The expansion of the required powers and factorials is done efficiently by iteration using the following relation for the multiplier, $t_i = \delta^i/i!$, of the ith derivative:

$$t_i = \frac{\delta^i}{i!} = \frac{\delta}{i} t_{i-1}.$$

Thus, each term of the sum requires just a constant multiplication by $1/i$ and two products of expansions, one by (δ/i) and the other by t_i.

The scalar and multivariate cases are equivalent in terms of computation. In the multivariate case, where either f or g is a vector or other non-scalar, the derivatives, f^i, are arrays and multiplication is not commutative. However, these derivatives are symmetric, which allows us to invoke the commutative law in computing δ^i. In both scalar and multivariate cases, the lists representing t_i may be sorted to reduce computation.

4.2.5 ROOTS OF EQUATIONS

If $f(g) = 0$, the root g may also be simply expanded using the above procedures. We assume that $f(g_0)$ is of order $n^{-1/2}$, a fact that will be clear from the integer partition approach. Further terms in the expansion may be computed iteratively by quasi-Newton iteration.

The usual Newton–Raphson algorithm for finding the root of an equation begins with an approximate root g_0 and iterates with

$$g_{i+1} = g_i - (f^{(1)}(g_i))^{-1})f(g_i). \tag{4.4}$$

The adaptation of this for expansions replaces the inverse $(f^{(1)}(g_i))^{-1}$ in the Newton–Raphson step with $(f_0^1)^{-1}$, the inverse of the leading term in its expansion. This modification will again be clear from the integer partition approach.

This leads to a surprisingly simple iteration where each Newton step adds precisely one additional term to the expansion. The procedure for computing the expansion of g, the root of the equation $f(g) = 0$, to order k, where g_0 is the leading term in the expansion, is as follows:

1. initialize: $g_1 = -(f_0^1)^{-1} f(g_0)$,
2. for $i = 2$ to k, do: $g_i = g_{i-1} - (f_0^1)^{-1} f(g_{i-1})$.

Each term of the expansion of the root requires just the expansion of a function and multiplication by a constant.

The interpretation of g_0 changes as we move from the usual Newton–Raphson algorithm to our symbolic version. For the former, g_0 is an initial guess for the true root while, for the latter, it is the leading term in the asymptotic expansion of the root, that is, its asymptotic value. For example, if g is a maximum likelihood estimator then g_0 would be the true value of the parameter. Similarly, if g were a Cornish–Fisher type expansion for a quantile, g_0 would be a quantile from the reference distribution.

4.2.6 MEANS

Many expansions will involve averages and products of averages with independent or dependent components. If X denotes a random variable with n components, the average may be expressed as

$$\overline{X} = \widetilde{X} + \dot{X},$$

where $\widetilde{X}$ denotes $\overline{E[X_i]}$, the average with components $E[X_i]$ and $\dot{X}$ denotes the average with components $(X_i - E[X_i])$.

If the components of X have finite mean and variance, $\widetilde{X}$ is of order $O(1)$ and $\dot{X}$ is $O_P(1/\sqrt{n})$. The expansion to order k is therefore the list of length $k + 1$,

$$\overline{X} \to \{\widetilde{X}, \dot{X}, 0, \ldots, 0\},$$

where the last $k - 1$ entries are all 0.

4.3   An integer partition approach

Expansions derived from other expansions can also be computed term by term where each term in the result depends on those of the component expansions. These terms may be identified uniquely by integer partitions which can increase efficiency and provide simple elegant insights into the structure of the terms. In some cases, efficiency is gained through the avoidance of redundant calculations. For example, in the quasi-Newton algorithm above, each step, $\hat{\delta} = (f_0^1)^{-1} f(g_{i-1})$, contributes only one additional term of order i. However, in the expansion of $f(g_{i-1})$ lower-order terms will be generated only to be cancelled. Since integer partitions identify only the contributing term, they avoid this redundant calculation.

However, for most common statistics, the methods of the previous section involve fewer calculations. Deriving an asymptotic expansion is generally regarded as technically complex, but integer partitions are something everyone can understand.

As an example, consider again the expansion of the product $t = g \times h$. The second term in the result,

$$t_2 = g_0 h_2 + g_1 h_1 + g_2 h_0,$$

is a sum over all partitions of the integer 2 that have length 2, namely

$$\{(0|2), (1|1), (2|0)\}.$$

All asymptotic expansions have this structure. All expansions are sums of products where the order of a product is the sum of the orders of each term in the product. So, a contribution to a term in an expansion can be identified by an integer partition. This idea is the basis for all the rules in this section. We identify rules for the expansion of products, powers, smooth functions, and their roots. In each case, the rule involves an integer partition. Algorithms that are recursive produce products that are identified by recursive integer partitions. Recursive integer partitions are graphs. In the multivariate case, the structure of a product of arrays is identified by the graph for the associated recursive integer partition.

4.3.1   DEFINITION OF INTEGER PARTITIONS

For a positive integer k, a partition $p = (b_1 | \cdots | b_j)$ of k is any set of integers $0 \le b_r \le k, r = 1, \ldots, j \le k$ such that $\sum_{r=1}^{j} b_r = k$. The collection of all such partitions of length j is denoted $\mathcal{P}_{j,k}$. A separate notation $\mathcal{P}^+$ is used when the restriction $b_r > 0$ applies. Finally, let $\mathcal{P}_k^+ = \cup_{k=1}^j \mathcal{P}_{j,k}^+$, $k > 0$. Some examples:

$$\mathcal{P}_{3,2} = \{(0|3), (3|0), (2|1), (1|2)\},$$

$$\mathcal{P}_{5,2}^+ = \{(1|4), (4|1), (3|2), (2|3)\},$$

$$\mathcal{P}_3^+ = \{(3), (2|1), (1|2), (1|1|1)\}.$$

In what follows, we construct a coefficient operator $\cdot_k$ that returns the kth term in the expansion of its argument. Rules for products powers, series, and roots are identified. Unless otherwise specified, we assume that $p = (b_1| \cdots |b_j)$.

4.3.2 EXPANSION OF PRODUCTS AND INTEGER POWERS

For the product $g \times h$ the kth term in the expansion has contributions:

Term		Subscripts
$g_0 h_k$	$\rightarrow$	$0, k$
$g_1 h_{k-1}$	$\rightarrow$	$1, k-1$
$g_2 h_{k-2}$	$\rightarrow$	$2, k-2$
$\cdots$	$\rightarrow$	$\cdots$
$g_k h_0$	$\rightarrow$	$k, 0$

where the pattern of subscripts identifies that the rule used also generates $\mathcal{P}_{2,k}$. If h is itself a product of $j-1$ expansions, this rule is iterated and the set $\mathcal{P}_{j,k}$ will result. In other words, the algorithm for the expansion of a product is the same algorithm used for generating $\mathcal{P}_{j,k}$. Thus, if we have expansions $g_i = g_{0,i} + g_{1,i} + g_{2,i} + \cdots$, $i = 1, \ldots, j$, we may write $\pi = \prod_{i=1}^j g_i$, where

$$\pi_k = \sum_{p \in \mathcal{P}_{j,k}} \prod_{i=1}^j g_{b_i, i}.$$

The case of integer powers is simpler because we can invoke the commutative law of multiplication. For instance, for the jth term in the expansion of the power g^2, the partitions $\{(i, j-i), (j-i, i)\}$ identify terms that make equivalent contributions to the result, namely $g_i g_{j-i}$ and $g_{j-i} g_i$. So, for any integer power only the sorted partition identifies distinct contributions. In such cases, sorting an integer partition provides the same efficiency that sorting lists did in Section 4.2.4. For multivariate Taylor series expansions, the use of sorted integer partitions elegantly invokes the symmetry of derivative arrays.

4.3.3 EXPANSION OF A SMOOTH FUNCTION OF AN EXPANSION

The Taylor series expansion of Section 4.2.4 is a sum of products of expansions. Hence, the term $f(g)_k$ can be identified by a set of integer partitions, as is evident from first few terms in the expansion:

	Term	Subscripts
	$f_0^0 \;\to$	$\{(0)\}$
	$f_1^0 + f_0^1 g_1 \;\to$	$\{(1), (0\|1)\}$
$f_2^0 + f_1^1 g_1 + f_0^1 g_2 + f_0^2 g_1^2/2 \;\to$		$\{(2), (1\|1), (0\|2), (0\|1\|1)\}$
$\vdots$	$\to$	$\vdots$

Since $(g - g_0)_0 = 0$, 0's can occur only in the first position of a partition. Thus, the general rule may be given in terms of $\mathcal{P}_{k+1}^+$:

$$f(g)_k = \sum_{p \in \mathcal{P}_{k+1}^+} f_{b_1-1}^{|p|-1} \prod_{i=2}^{|p|} g_{b_i} / (|p| - 1)!.$$

Use of the set $\mathcal{P}_{k+1}^+$ implicitly invokes the fact that $(g - g_0)_0 = 0$.

Because derivatives are symmetric, the partition may be sorted holding the first element fixed. This simplifies the expression.

4.3.4 ROOTS OF EQUATIONS

When g is the root of the equation $f(g) = 0$, terms in its expansion may be found by solving $f(g)_k = 0$ for g_k. The first few equations are

$$0 = f_0^0,$$
$$0 = f_1^0 + f_0^1 g_1,$$
$$0 = f_2^0 + f_1^1 g_1 + f_0^1 g_2 + f_0^2 g_1^2/2,$$
$$\vdots$$

where the first of these gives the condition mentioned earlier that $f(g_0)$ is of order $o(1)$. Premultiplication by $-(f_0^1)^{-1}$, the aforementioned modification of the Newton–Raphson algorithm, is clear from solving the remaining equations:

$$g_1 = -(f_0^1)^{-1} f_0^1$$
$$g_2 = -(f_0^1)^{-1}(f_2^0 + f_1^1 g_1 + f_0^2 g_1^2/2)$$
$$\vdots$$

Since the right-hand side of each equation depends on $\mathcal{P}_{k+1}^+$, the solution will as well, although the partition, $(1|k)$, identifying the term $f_0^1 g_k$ is omitted. Letting

$\mathcal{P}_{k+1}^{+} - (1|k)$ denote the set $\mathcal{P}_{k+1}^{+}$ with the partition $(1|k)$ dropped, the general result is

$$g_k = \sum_{p \in \{\mathcal{P}_{k+1}^{+} - (1|k)\}} \tilde{f}_{b_1-1}^{|p|-1} \prod_{i=2}^{|p|} g_{b_i} / (|p| - 1)!, \qquad (4.5)$$

where $\tilde{f}_i^r = (-f_1^0)^{-1} f_i^r$. Note that this rule is remarkably similar to the series rule. This rule is recursive and terms generated will be identified by recursive integer partitions.

4.3.5 IMPLEMENTATION

In a manner similar to the previous chapter, integer partitions are returned as sums. The partition of an integer k into j blocks is computed recursively as

$$\mathcal{P}_{1,k} = \{k\},$$
$$\mathcal{P}_{j,k} = Q(\mathcal{P}_{j-1,k}),$$

where Q is defined as a linear operator such that

$$Q(\{b_1, \ldots, b_r\}) = \sum_{i=0}^{b_r} \{b_1, \ldots, b_r - i, i\}.$$

For $k = 2$, note the similarity between the rule

$$Q(\{k\}) = \sum_{i=0}^{k} \{i, k - i\}.$$

and the kth term in the expansion of $g \times h$, namely $\sum_{i=0}^{k} g_i h_{k-i}$. It is at this point that the use of integer partitions can again be appreciated. Many expansions will rely on the same integer partition and efficiency is gained if it is computed only once. The partition operator $\mathcal{P}^{+}$ is defined in a similar fashion, but the lower limits in the above sums equal 1.

A dot product Following the conventions of the previous chapter, products have list representations. For example

$$\pi = \prod_{i=1}^{j} g_i \ \rightarrow \ \omega_1 = \{g_1, \ldots, g_j\}.$$

Most computer algebra environments allow a general version of a dot product operator. Here we define the operator $\cdot$ to return the term identified by a list of variables and a partition. For example,

$$\{X, Y, Z\} \cdot \{3, 0, 1\} \rightarrow X_3 Y_0 Z_1,$$

where the operator $\cdot$ may return an explicit product or the list that represents that product. Note the difference between this and the usual convention for $\cdot$ which would return $3X + Z$.

Defining $\cdot$ to be a linear operator the rule for expansion of a product is simply

$$\pi_k = \omega_1 \cdot \mathcal{P}_{j,k},$$

where ω_1 is defined above and where $(g)_k$ is an operator that returns the kth coefficient of its argument. The implementation of the rules for powers, series, and roots are similar and left as exercises.

4.4 Multivariate expansions

Multivariate expansions involve objects of more than one dimension and are typically viewed as vastly more complicated than scalar expansions. This is false. If we follow the conventions of the previous chapter, the procedures defined here will return products represented as lists. Sorting these lists mimics the commutative rule for scalar multiplication. In the multivariate case, multiplication does not commute and these lists remain unsorted. In other words, automating expansions in the multivariate case is accomplished when the procedures simply return unsorted lists. They involve *fewer* procedures.

In addition, for expansions involving series, the dimension of an array is given by the length of a list. This is especially useful when rules are recursive. Here nested lists identify the dimension of all arrays involved as well as the structure of inner products.

4.5 A likelihood example

For a sample $y_1, \ldots, y_n$ with density f_θ, the likelihood function

$$L(\theta) = f_\theta(y_1, \ldots, y_n)$$

gives a measure of 'consistency with the data' for values of θ. Typically one considers the log-likelihood

$$l(\theta) = \log\{L(\theta)\} = \sum_{i=1}^{n} l_i(\theta),$$

which is a sum and hence more convenient to study. Here $l_i(\theta) = \log\{f_\theta(y_i)\}$ is the contribution of the ith data point, y_i. Natural quantities of interest associated with $l(\theta)$ are its maximum $l(\hat{\theta})$ and the distance $w(\theta) = 2\{l(\hat{\theta}) - l(\theta)\}$ from this maximum for a value θ. The point $\hat{\theta}$ is called the maximum likelihood estimate and $w(\theta)$ the log-likelihood ratio statistic. If $\hat{\theta}$ can be obtained by solving $\partial_\theta l(\theta) = 0$, asymptotic expansions for $\hat{\theta}$ and $w(\theta)$ can be obtained using the algorithms of this chapter.

Since they are sums, the log-likelihood function and its derivatives may be expressed as averages:

$$l(\theta) = n\bar{l},$$

$$l^{(r)}(\theta) = \frac{\partial^r}{\partial \theta^r} l(\theta) = \sum_{i=1}^{n} \frac{\partial^r}{\partial \theta^r} l_i(\theta) = n\bar{l}_r.$$

These averages are convenient for expansions because they lead to simpler expressions, devoid of powers of n. Using Section 4.2.6, these can be expanded. For example,

$$\bar{l}_2 \to \{\tilde{l}_2, \dot{l}_2, 0, \ldots, 0\},$$

where the argument θ is suppressed and the last $k - 2$ terms in the expansion are again 0.

The maximum likelihood estimate $\hat{\theta}$ is the root of the equation $\overline{l^{(1)}(\hat{\theta})} = 0$. Applying the procedures of this chapter yields the expansion for $\hat{\theta}$ displayed for $k = 3$ in the more general, multivariate, non-identically distributed case:

$$
\begin{aligned}
(\theta) &+ \left(-\tilde{l}_2^{-1} \cdot \dot{l}_1 \right) \\
&+ \left(\tilde{l}_2^{-1} \cdot \dot{l}_2 \cdot \tilde{l}_2^{-1} \cdot \dot{l}_1 - \tfrac{1}{2}\tilde{l}_2^{-1} \cdot \tilde{l}_3 \cdot \tilde{l}_2^{-1} \cdot \dot{l}_1 \cdot \tilde{l}_2^{-1} \cdot \dot{l}_1 \right) \\
&+ \left(-\tilde{l}_2^{-1} \cdot \dot{l}_2 \cdot \tilde{l}_2^{-1} \cdot \dot{l}_2 \cdot \tilde{l}_2^{-1} \cdot \dot{l}_1 \right. \\
&\quad - \tfrac{1}{2}\tilde{l}_2^{-1} \cdot \dot{l}^3 \cdot \tilde{l}_2^{-1} \cdot \dot{l}_1 \cdot \tilde{l}_2^{-1} \cdot \dot{l}_1 \\
&\quad + \tfrac{1}{2}\tilde{l}_2^{-1} \cdot \dot{l}_2 \cdot \tilde{l}_2^{-1} \cdot \tilde{l}_3 \cdot \tilde{l}_2^{-1} \cdot \dot{l}_1 \cdot \tilde{l}_2^{-1} \cdot \dot{l}_1 \\
&\quad + \tilde{l}_2^{-1} \cdot \tilde{l}_3 \cdot \tilde{l}_2^{-1} \cdot \dot{l}_1 \cdot \tilde{l}_2^{-1} \cdot \dot{l}_2 \cdot \tilde{l}_2^{-1} \cdot \dot{l}_1 \\
&\quad + \tfrac{1}{6}\tilde{l}_2^{-1} \cdot \tilde{l}_4 \cdot \tilde{l}_2^{-1} \cdot \dot{l}_1 \cdot \tilde{l}_2^{-1} \cdot \dot{l}_1 \cdot \tilde{l}_2^{-1} \cdot \dot{l}_1 \\
&\quad \left. - \tfrac{1}{2}\tilde{l}_2^{-1} \cdot \tilde{l}_3 \cdot \tilde{l}_2^{-1} \cdot \dot{l}_1 \cdot \tilde{l}_2^{-1} \cdot \tilde{l}_3 \cdot \tilde{l}_2^{-1} \cdot \dot{l}_1 \cdot \tilde{l}_2^{-1} \cdot \dot{l}_1 \right),
\end{aligned}
\tag{4.6}
$$

where, recall, $\tilde{l}_r$ is used to denote $\overline{E[l_i^{(r)}(\theta)]}$, the average of the rth derivative of the components of the likelihood.

Array products are denoted by $\cdot$, the usual dot product operator. The above form avoids the all too common subscripts and superscripts. Terms of the same order are grouped with parentheses.

The expansion of the likelihood ratio statistic divided by n, $\overline{w(\theta)} = w(\theta)/n$, is computed for $k = 4$ as

$$(0) + (0) - \left(l_1 \cdot \tilde{l}_2^{-1} \cdot l_1 \right)$$

$$+ \left(l_2 \cdot \tilde{l}_2^{-1} \cdot l_1 \cdot \tilde{l}_2^{-1} \cdot l_1 - \frac{1}{3}\tilde{l}_3 \cdot \tilde{l}_2^{-1} \cdot l_1 \cdot \tilde{l}_2^{-1} \cdot l_1 \cdot \tilde{l}_2^{-1} \cdot l_1 \right)$$

$$+ \left(- l_2 \cdot \tilde{l}_2^{-1} \cdot l_1 \cdot \tilde{l}_2^{-1} \cdot l_2 \cdot \tilde{l}_2^{-1} \cdot l_1 \right.$$

$$- \frac{1}{3} l_3 \cdot \tilde{l}_2^{-1} \cdot l_1 \cdot \tilde{l}_2^{-1} \cdot l_1 \cdot \tilde{l}_2^{-1} \cdot l_1$$

$$+ \tilde{l}_3 \cdot \tilde{l}_2^{-1} \cdot l_1 \cdot \tilde{l}_2^{-1} \cdot l_1 \cdot \tilde{l}_2^{-1} \cdot l_2 \cdot \tilde{l}_2^{-1} \cdot l_1$$

$$+ \frac{1}{12}\tilde{l}_4 \cdot \tilde{l}_2^{-1} \cdot l_1 \cdot \tilde{l}_2^{-1} \cdot l_1 \cdot \tilde{l}_2^{-1} \cdot l_1 \cdot \tilde{l}_2^{-1} \cdot l_1$$

$$\left. - \frac{1}{4}\tilde{l}_3 \cdot \tilde{l}_2^{-1} \cdot l_1 \cdot \tilde{l}_2^{-1} \cdot l_1 \cdot \tilde{l}_2^{-1} \cdot \tilde{l}_3 \cdot \tilde{l}_2^{-1} \cdot l_1 \cdot \tilde{l}_2^{-1} \cdot l_1 \right).$$

Terms in the expansion of $w(\theta)$ are easily obtained from the above expansion. In the identically distributed case, or the scalar case, further simplifications of the above expressions result.

4.6 A notational convention

The above results are expressed in a compact notation. However, for historical reasons, much of the statistical literature uses a more complex notation. The use of a notational filter allows the above results to be presented using a standard notation.

In the multivariate case, it is common to use the summation convention to identify array products in an expansion. Here arrays are represented componentwise by using subscripts and superscripts. For example, a $(k \times 1)$-dimensional vector v and a $(k \times l)$-dimensional matrix M are represented as v_r and M_{rs}, respectively, where $r = 1, \ldots, k, s = 1, \ldots, l$.

4.6.1 DIMENSIONS OF ARRAYS

If the parameter θ has p components, the derivatives in the above expansions for $\hat{\theta}$ and $\overline{w(\theta)}$ are arrays of dimensions depending on p. The first derivative, $l^{(1)}(\theta) = \delta l(\theta)/\delta\theta_i$, $1 \le i \le p$, is a one-dimensional vector with p components. The second derivative, $l^{(2)}(\theta) = \delta^2 l(\theta)/(\delta\theta_i \delta\theta_j)$, is a matrix with dimensions $p \times p$. It is two dimensional. In general, the kth derivative, $l^{(k)}(\theta)$ is an array of size $p \times p \cdots \times p$. It has k dimensions. The trivial procedure for computing the dimension of a term is implemented by

$$\dim(l^{(k)}(\theta)) = k.$$

All the arrays involved in the expansions of $\hat{\theta}$ and of $w(\theta)$ have dimensions of the same length. Let p denote this common length.

4.6.2 NOTATION FOR PRODUCTS OF ARRAYS

The product $M \cdot v$ is a vector with p components. The ith component is

$$(M \cdot v)_i = \sum_{j=1}^{p} M_{ij} v_j.$$

If the length of each dimension of all arrays is the same, as in the case of pseudo-likelihood expressions, the product may be more simply denoted by

$$(M \cdot v)_i = M_{ij} v_j,$$

where a repetition of an index indicates summation over the p values of that index.

4.6.3 PROCEDURE FOR INDEXING PRODUCTS OF ARRAYS

In this notation, the dot product of two arrays is represented by a product of components from the two terms with the last index of the first term being the same as the first index of the second.

Array multiplication is associative:

$$A \cdot B \cdot C = (A \cdot B) \cdot C.$$

Therefore, the indexing of a multiple product may be defined recursively, starting with the first factor. This leads to the following procedure for representing the product $A \cdot B \cdot C \cdots$.

The procedure is described in terms of lists of indices. For example, here we use members of the list $\mathcal{I} = \{r, s, t, u, v, w, x, y, z\}$ as in, for example, McCullagh (1987). Let $\{u_1, \ldots u_j\}$ denote the indices used but not repeated in previous terms, and let $\{f_1, \ldots f_k\}$ be the list of free indices, indices from the possible list which have not yet been used. Then the indexing function is defined by

$$\mathcal{N}[A \cdot B \cdots, \{u_1, \ldots u_j\}, \{f_1, \ldots f_k\}]$$

$$= A_{u_j, f_1, \ldots f_{\dim[A]-1}}$$

$$\times \mathcal{N}[B \cdots, \{u_1, \ldots u_{j-1}, f_1, f_{\dim[A]-1}\}, \{f_{\dim[A]}, f_{\dim[A]+1}, \ldots f_k\}]$$

and, if no indices are used only once,

$$\mathcal{N}[A \cdot B \cdots, \{\}, \{f_1, \ldots f_k\}]$$

$$= A_{u_j, f_1, \ldots f_{\dim[A]}} \times \mathcal{N}[B \cdots, \{f_1, \ldots f_{\dim[A]}\}, \{f_{\dim[A]+1}, f_{\dim[A]+2}, \ldots f_k\}].$$

With this recursive function defined, the product is then indexed by $\mathcal{N}[A \cdot B \cdots, \{\}, \mathcal{I}]$:

$$\tilde{l}_3 \longrightarrow I_{rst},$$

$$\dot{l}_2 \longrightarrow z_{rs},$$

$$\tilde{l}_2^{-1} \cdot \dot{l}_2 \cdot \tilde{l}_2^{-1} \cdot \dot{l}_1 \longrightarrow K^{rs} z_{st} K^{tu} z_u,$$

where $K^{rs} = \{I_{rs}\}^{-1}$. The above expansion for $\hat{\theta}$ may be written in this notation using

$$\mathcal{N}[\hat{\theta}, \{\}, \mathcal{I}] \longrightarrow \left(- z_s\, K^{rs}\right)$$
$$+ \left(z_{st}\, z_u\, K^{rs}\, K^{tu} - \tfrac{1}{2} I_{stu}\, z_v\, z_f\, K^{rs}\, K^{tv}\, K^{uw}\right)$$
$$+ \left(- \tfrac{1}{2} z_{stu}\, z_v\, z_w\, K^{rs}\, K^{tv}\, K^{uw} - z_{st}\, z_{uv}\, z_w\, K^{rs}\, K^{tu}\, K^{vw}\right.$$
$$+ \tfrac{1}{6} I_{stuv}\, z_w\, z_x\, z_y\, K^{rs}\, K^{tw}\, K^{ux}\, K^{vy}$$
$$+ \tfrac{1}{2} I_{uvw}\, z_{st}\, z_x\, z_y\, K^{rs}\, K^{tu}\, K^{vx}\, K^{wy}$$
$$+ I_{stu}\, z_v\, z_{wx}\, z_y\, K^{rs}\, K^{tv}\, K^{uw}\, K^{xy}$$
$$\left. - \tfrac{1}{2} I_{stu}\, I_{wxy}\, z_v\, z_z\, z_i\, K^{rs}\, K^{tv}\, K^{uw}\, K^{xz}\, K^{yi}\right).$$

The standard asymptotic $N(0, -K^{r,s}/n)$ distribution for $\hat{\theta}$ is evident from the first term in the expansion. The remaining terms contribute to various departures from this approximation. This expression appears in McCullagh (1987). The fourth-order term in the expansion has 22 terms while the fifth has 82. These are not printed here.

Writing the expansion for $\overline{w(\theta)}$ in this notation yields

$$w(\theta) \longrightarrow \left(- z_r\, z_s\, K^{rs}\right) + \left(z_{rs}\, z_t\, z_u\, K^{rt}\, K^{su} - \frac{1}{3} I_{rst}\, z_u\, z_v\, z_w\, K^{ru}\, K^{sv}\, K^{tw}\right)$$
$$\times \left(- \frac{1}{3}\left(z_{rst}\, z_u\, z_v\, z_w\, K^{ru}\, K^{sv}\, K^{tw}\right)\right.$$
$$+ \frac{1}{12} I_{rstu}\, z_v\, z_w\, z_x\, z_y\, K^{rv}\, K^{sw}\, K^{tx}\, K^{uy}$$
$$- z_{rs}\, z_t\, z_{uv}\, z_w\, K^{rt}\, K^{su}\, K^{vw}$$
$$+ I_{rst}\, z_u\, z_v\, z_{wx}\, z_y\, K^{ru}\, K^{sv}\, K^{tw}\, K^{xy}$$
$$\left. - \frac{1}{4} I_{rst}\, I_{wxy}\, z_u\, z_v\, z_z\, z_i\, K^{ru}\, K^{sv}\, K^{tw}\, K^{xz}\, K^{yi}\right).$$

The standard asymptotic χ^2_p distribution for $w(\theta)$ is again evident from the first term in the expansion. The remaining terms contribute to various departures from this approximation. This expression appears in McCullagh (1987). The third-order term has 14 terms while the fourth has 58. These are not printed here.

4.7 Canonical form, integer partitions, and graphs

Manipulating objects that directly use the summation convention results in the difficulty of getting the computer to recognize identical terms. For example, the computer will not recognize that $v^r M_{rs}$ and $v^t M_{ts}$ are the same, even though *we* know replacing the index r with t does not affect the value of $v^r M_{rs}$. Terms involving many indices make this problem considerably worse and can lead to a large number of equivalent terms in a final result. Recognition of this equivalence can simplify the expression.

A simple procedure for reducing each term to its canonical form is to compute the set of expressions that result from permuting the repeated indices. The canonical form is the first term in the lexicographical ordering of this set. For example,

$$A_s \cdot B_{sr} \cdot C_r \overset{\text{permute}}{\rightarrow} \{A_s \cdot B_{sr} \cdot C_r, A_r \cdot B_{rs} \cdot C_s\}$$
$$\overset{\text{sort}}{\rightarrow} \{A_r \cdot B_{rs} \cdot C_s, A_s \cdot B_{sr} \cdot C_r\}$$
$$\overset{\text{first}}{\rightarrow} A_r \cdot B_{rs} \cdot C_s.$$

This procedure is simple to implement and is described in detail in Chapter 7. More efficient procedures are described in Chapter 10.

The integer partitions are useful because they provide an index-free representation of products of arrays. To demonstrate this, consider a term that contributes to the expansion of $\hat{\theta}$ to fifth order,

$$T = K^{ab} I_{bcd} K^{ce} K^{df} I_{egh} I_{fij} K^{gk} K^{hl} K^{im} K^{jn} z_k z_l z_m I_{nop} K^{oq} K^{pr} z_q z_r.$$

Since the symbolic version of the Newton–Raphson algorithm is recursive, terms in the expansion of the root of a function are identified by recursive integer partitions. The reader may greet this with scepticism as it is not immediately apparent what integer partition identifies T. However, T has a graphical representation that coincides with a unique recursive integer partition and *vice versa* (Figure 4.1). Ignoring the index a (it is not a duplicated index and does not indicate an array product but rather the fact that T is a vector) and tracing the inner product from b through c and d and eventually to r, we draw a graph in which a line in the graph connects quantities that share an index.

The important feature of this graph is the splits and, by 'ignoring' the Ks, we can simplify the representation. Note that this is equivalent to the notational convention for $\tilde{f}^j = (-f_0^1)^{-1} f^j$ and essentially makes the denominator in the Newton–Raphson step implicit. The expression becomes

$$\tilde{I}_{abc} \tilde{I}_{bde} \tilde{I}_{cfg} \tilde{z}_d \tilde{z}_e \tilde{z}_f \tilde{I}_{ghi} \tilde{z}_h \tilde{z}_i,$$

where $\tilde{I}_{i\ldots} = K^{ij} I_{j\ldots}$ and $\tilde{z}_{i\ldots} = K^{ij} z_{j\ldots}$. Now, when we draw a graph, it gives a recursive partition of the integer 5. In other words, recursive integer partitions are graphs that capture the structure of an array product.

4.8 Bibliographical notes

Aspin (1948) uses an iterative technique that subsequently appears in a series of papers (Welch and Peers, 1963; Welch, 1965; Peers and Iqbal, 1985; DiCiccio, Hall and Romano, 1991) where an expansion for the root of some function is required for study. The notion of using integer partitions to identify terms is found in several

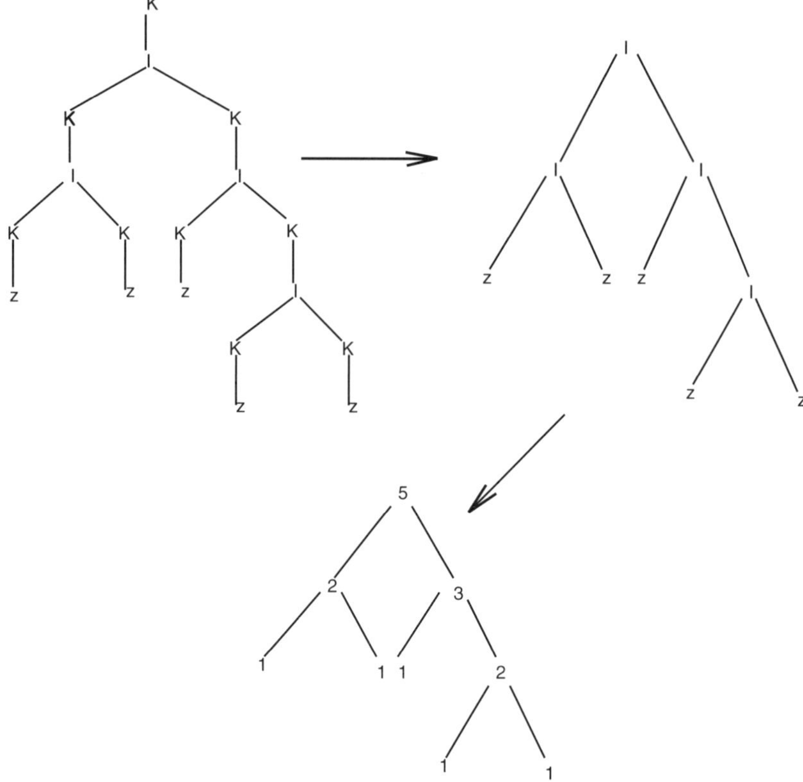

Fig. 4.1. One of the 109 terms that arise in the expansion of the maximum likelihood estimate to order 5.

places in McCullagh (1987). Their use to identify terms in an asymptotic expansion in an automatic way is given in Stafford (2000). Venables (1985) uses symbolic computation to obtain expansions for maximum marginal likelihood estimates.

4.9 Problems

1. Compute the expansion of the estimate of variance $\widehat{\sigma^2}$ and use this to compute the expansion of $\hat{\sigma}$.

2. Using the above, compute the expansion of $t = (\overline{X} - E[X])/\hat{\sigma}$.

3. Compute the expansion of the mode of the posterior density function $p(\theta|y) \propto p(\theta)l(\theta)$, where $p(\theta)$ is the prior and $l(\theta)$ is the likelihood. Note that the method is similar to the case of the maximum liklihood estimate except that there is an additional leading term for each derivative of $\log(p(\theta|y))$.

4. Use the methods of this chapter to compute the expansion of the signed root $\sqrt{l(\hat{\theta})}$ when θ is scalar.

5. Compute the expansion of $g(\hat{\theta})$, a smooth function of the maximum likelihood estimate. How does this expansion differ from the expansion for $l(\hat{\theta})$?

6. Compute the expansion of $g(\hat{\theta})$, a smooth function of an M-estimate. How does this expansion differ from the expansion of $g(\hat{\theta})$ in the previous question?

7. Compute the expansion of $\hat{i} = -l^{(2)}(\hat{\theta})/n$, the observed information. Use this to compute expansions for the Wald, $\sqrt{n}(\hat{\theta} - \theta)\hat{i}^{1/2}$, and Rao, $l(\theta)\hat{i}^{-1/2}/\sqrt{n}$, statistics.

5
Expansions of expectations, cumulants, and unbiased estimates

5.1 Introduction

The remaining chapters of this book present applications of the core operators to various statistical methods involving cumulants or cumulant expansions. Expansions of cumulants may be computed using the operators of the previous two chapters. Matters are complicated by the fact that the expected values have expansions with differing powers of n. These, in turn, contribute to *several* different terms in the resulting expansion. For example, consider the second moment of a mean of IID random variables,

$$E[\overline{X}^2] \longrightarrow E[X]^2 + \frac{E[X^2]}{n} - \frac{E[X]^2}{n}, \tag{5.1}$$

with one term of order $O(1)$ and two terms of order $O(n^{-1})$.

Additional rules are needed for the expansion of expectations.

5.2 Expansions of expectations

Chapter 4 presented procedures for computing expansions of many common statistics. Terms of these expansions involve products of averages, quantities represented in Chapter 3 as type A. The expected value of such quantities may be computed using the operator $\mathcal{T}_{\mu,\Delta[\mu],\Delta,A}$, if the components of the average are IID random variables, and by using $\mathcal{T}_{A[\mu],\Delta[\mu],\Delta,A}$, if they are not identically distributed. Both of these involve a coefficient transformation that generates constants expressed in powers of n^{-1}—the determining criterion of the order in the expansion of an expectation. Since these constants are of order $\leq O(1)$, the expectation of any product of averages of order $O(n^{-k})$ results in terms of order $O(n^{-(k+j)})$ where $j \geq 0$. Therefore, to compute the expectation $E[g]$ to $O(n^{-k})$, we need only expand the argument, g, to that order. For example, to produce (5.1) to order $O(n^{-1})$, we apply the expectation operator to the expansion of $\overline{X}^2$ computed to the same order. The resulting terms are assigned to the appropriate positions in the list representing the expansion

$$E[\overline{X}^2] \to \{E[X]^2,\ 0,\ E[X^2]/n + E[X]^2/n\},$$

where the appropriate positions themselves are determined by the order.

5.3 Expansion of cumulants

Recall from Chapter 3 that a cumulant

$$\kappa_{\hat{\theta}_1,\hat{\theta}_2,\ldots} = \kappa_V \tag{5.2}$$

may be considered as a function of a set V, where $V = \{\hat{\theta}_1, \hat{\theta}_2, \ldots\}$. The cumulant may be written in terms of moments using the transformation $\mathcal{T}_{\mu,\kappa}$ which implements the identity (3.5),

$$\kappa_V = \sum_{p \in \mathcal{P}_V} c^p \prod_{b \in p} \mu_b, \tag{5.3}$$

where b represents a product of a subset of variables.

If the variables are represented by expansions, these products are computed automatically using the multiplication procedure of Chapter 4. The factors μ_b, therefore denote the expectation of an expansion which is computed as in Section 5.2 and so the μ_b are themselves expansions. Once again, the expansion of the product $\prod_{b \in p} \mu_b$ is automatically computed using the multiplication procedure of Chapter 4. Therefore, the expansion of a cumulant $\kappa_{\hat{\theta}_1,\ldots,\hat{\theta}_k}$ is given by the algorithm

1. create the list $V = \{\hat{\theta}_1, \ldots, \hat{\theta}_k\}$;
2. compute W by applying $\mathcal{T}_{\mu,\kappa}$ to V to express the cumulant as products of expectations;
3. expand the products of $\hat{\theta}_i$ as expansions;
4. compute $X = \mathcal{T}_{\Delta,A} \circ W$;
5. compute $Y = \mathcal{T}_{A[\mu],\Delta[\mu],\Delta} \circ X$;
6. expand the products of expectations.

Note that, depending on the implementation, steps 3 and 6 may be automatic.

This approach is similar to the spirit of Section 3.11.3, where cumulants of statistics involving products of averages such as k-statistics, were represented and transformed to expressions involving products of averages of moments using the operator $\mathcal{T}_{\mu,\kappa[A]}$.

5.4 Efficient computation of expansions of cumulants

The procedure described above will result in much unnecessary computation if implemented without modification. The following sections outline steps that will eliminate this.

5.4.1 EXPECTATIONS OF STANDARDIZED MEANS

The expansion of a mean $\overline{X}$ was defined in Chapter 4 as

$$\overline{X} = \widetilde{X} + \dot{X},$$

where $\widetilde{X} = \overline{E[X]}$ and the term $\dot{X}$ is random of order $O_P(n^{-1/2})$ with expectation 0. In general, a product of moments equals 0 if any factor is the expectation of a centred average and hence 0.

Recall from Section 5.2 that terms of expansions involve products of centred averages (type A). Expectation of these terms is automated using either $\mathcal{T}_{\mu,\Delta[\mu],\Delta,A}$ or $\mathcal{T}_{A[\mu],\Delta[\mu],\Delta,A}$. Both transformations involve $\mathcal{T}_{\Delta,A}$. Use of this transformation results in applying the partition operator $\mathcal{P}$ to a list of type A. For the resulting partitions, blocks with only one centred average will evaluate to expectation zero upon application of $\mathcal{T}_{\Delta[\mu],\Delta}$. Hence, these partitions are immediately set to zero and computing the expected value of other blocks in the partition is avoided.

5.4.2   VANISHING LEADING TERMS

Some expansions may have leading terms equal to zero and efficiency can be gained if this is accounted for. Consider the expansion to order $O(n^{-k})$ of the product ab, both of which may be expectations. In general, the expansions of both a and b to order $O(n^{-k})$ are required. However, if the leading term of one expansion is 0, then the expansion of the *other* factor is required only to order $O(n^{-k-1/2})$.

An example of such a case is when one factor in a product is the expectation of a product of centred averages such as $E[\dot{X}^3]$ whose expansion has four leading 0's. To avoid needless computation, it is useful to account for known leading 0's. This strategy may be implemented in a general way. The expectation of a product of k centred variables ($\dot{X}$'s) has order $\leq n^{-k/2}$. This order is simply computed by counting the number of $\dot{X}$'s. Now consider the product $\Pi_{j=1}^{m}\mu_j$ of moments, μ_j, with orders $n^{-k_j/2}$. The product has order $n^{-K/2}$, where $K = \sum^m k_j$. Suppose now that this product is to be expanded to order $n^{-J/2}$. Clearly if $K > J$ the result is 0. Otherwise, it follows that μ_j is multiplied by a product of order $n^{-\sum_{i\neq j}k_i/2} = n^{-(K-k_j)/2}$. As a result, the moment μ_j needs to be computed only to order $n^{-(J-(K-k_j))/2}$ and much computation can be avoided.

5.4.3   REQUIRED ORDER OF ARGUMENTS

Finally, higher cumulants depend only on *lower*-order terms of expansions of their arguments. Consider the expansion

$$g = g_0 + \cdots + g_k$$

whose leading term is constant. All cumulants above the first are invariant to addition of a constant and so, without loss of generality, we may assume $g_0 = 0$. This reduces computation. For example, expanding the variance of g to order $O(n^{-1})$ requires an expansion of g only to order $O(n^{-1/2})$.

In general, a cumulant involving $g = g_0 + \cdots + g_m$ could be computed from the $O(n^{1/2-m})$ expansion of $\sqrt{n}g = \sqrt{n}g_1 + \cdots + \sqrt{n}g_j$. The order arguments used in Section 5.2 indicate that a cumulant involving k such expansions would be computed correctly to order $O(n^{1/2-j/2})$. This cumulant, however, has been multiplied by $n^{k/2}$. It follows, therefore, that the cumulant involving the original arguments is correct to order $n^{1/2-j/2-k/2}$. A cumulant computed to order $n^{-m/2}$, therefore, requires expansion of its arguments only to order $n^{-j/2}$ where

$$j = m + 1 - k.$$

For example, a fourth-order cumulant ($k = 4$) computed to order $n^{-6/2}$ ($m = 6$) requires only expansions of its arguments to order $n^{-(6+1-4)/2} = n^{-3/2}$.

5.4.4 SUMMARY: CUMULANT EXPANSIONS

The computation of higher cumulants is simplified by the following modifications to the algorithm of Section 5.3:

- set the leading, constant term of order 1 to 0 in each argument in step 1 (Section 5.4.3);
- compute arguments of the cumulant only to the order required in step 1 (Section 5.4.3);
- set the expectations of all products of moments that involve a factor of the form $E[\dot{X}]$ to 0 without computing the other moments after step 4 (Section 5.4.1);
- accounting for leading vanishing terms, compute moments of the remaining products only to the required order in step 5 (Section 5.4.2).

These procedures are implemented with the transformation $\mathcal{T}_{\mu,\kappa[A]}$, or $\mathcal{T}_{A[\mu],\kappa[A]}$, which applied to a list of expansions $\{\hat{\theta}_1, \dots, \hat{\theta}_k\}$, defined by (5.2), computes the expansion of $\kappa_{\hat{\theta}_1,\dots,\hat{\theta}_k}$.

5.5 Exploiting integer partitions

Integer partitions were used in Section 4.3 to compute terms g_j of the expansion of g. The implementation causes a subscript j on an object to execute appropriate computations. For example, the second term in the expansion of the product ab may be computed as

$$(ab)_2 \to a_0\,b_2 + a_1\,b_1 + a_2\,b_0.$$

It may deepen the understanding of the previous sections of this chapter if one considers cumulant expansions in terms of the subscript operator $(\cdot)_j$. Hence, we offer an argument that roughly parallels Section 5.4.2.

For the set of arguments $V = \{X, Y, \dots\}$ the jth coefficient in the expansion of the cumulant is $\kappa_{V,j}$. This may be expressed alternatively in terms of moments by $(\mathcal{T}_{\mu,\kappa} \circ \kappa_V)_j$. For example,

$$(f_\mu \circ \mathcal{T}_{\mu,\kappa} \circ \kappa_{\{X,Y\}})_2 \longrightarrow E[XY]_2 - E[X]_0 E[Y]_2$$
$$- E[X]_2 E[Y]_0 - E[X]_1 E[X]_1$$

Therefore, the task is to compute the coefficients $E[g]_j$ from the argument g. As noted in Section 5.2,

$$E[g]_j \neq E[g_j]$$

and the rule is that

$$E[g]_j = \sum_{i=0}^{j} E[g_i]_{j-i}$$

must be used. However, this can lead to redundant computation if coefficients of order $k < j$ are also computed. The issue is mute if one considers the derivation of an expansion as a whole rather than a coefficient. While the rule $E[g]_j = E[g_j]$ is incorrect, its use gives the correct result for a truncated expansion: since the order of $E[g_j]$ is $\leq O(n^{-j/2})$, it follows that

$$\mathcal{AE}_i[E[g]] = \sum_{j=0}^{i} E[g]_j$$

$$= \sum_{j=0}^{i} E[g_j] + O(n^{-i+1})$$

$$= E[\mathcal{AE}_i[g]] + O(n^{-i+1}).$$

The result simply states that truncating $E[\mathcal{AE}_i[g]]$ gives $\mathcal{AE}_i[E[g]]$. This immediately extends to products of moments, and hence cumulants, since for $\prod_r E[g^r]$ we have

$$\mathcal{AE}_i\left[\prod_r E[g^r]\right] = \prod_r \mathcal{AE}_i[E[g^r]] + O(n^{-i+1})$$

$$= \prod_r E[\mathcal{AE}_i[g^r]] + O(n^{-i+1}).$$

Here g^r denotes a distinct expansion rather than a derivative as in Chapter 4.

Finally, the gains in efficiency from considering the standardized variable $\bar{g} = \sqrt{n}(g - g_0)$ are automatic if we define the rule

$$\bar{g}_j = g_{j+1}$$

and compute cumulants for $\bar{g}$ instead of g. For example, the expansion of the variance of g to order $O(n^{-1})$ is computed as

$$\kappa_{\{\bar{g},\bar{g}\},0} \longrightarrow E(g_1^2) - E(g_1)^2$$

and, as desired, only the expansion of g to order $O(n^{-1/2})$ is required.

5.6 A smooth function model

Often, a parameter g may be defined in terms of expectations as $g = g(\theta)$, where $\theta = E_F(X)$ and g is some smooth function. An estimate of g, $\hat{g}$, may be defined in terms of averages as $\hat{g} = g(\hat{\theta})$, where $\hat{\theta} = E_{F_n}(X)$. Here X is a random variable, possibly vector-valued. Means, variances, correlations, regression coefficients, and coefficients of variation are examples of parameters defined explicitly in terms of expectations. Given a score-type function Ψ, parameters associated with maximum-likelihood-type procedures, M-estimates, may be defined implicitly in terms of expectations by $E_F[\psi(X, \theta)] = 0$ with estimates defined by $E_{F_n}[\psi(X, \theta)] = 0$. Smooth functions of these may also be of interest.

Moving from parameters to estimates involves replacing the true distribution F with the empirical distribution F_n in the definition of expectation. The empirical distribution is a discrete uniform distribution with points of support given by a sample $X_1, \ldots, X_n \sim F$. Hence, moments of F_n are averages and estimates are therefore defined explicitly or implicitly in terms of averages. Efron and Tibshirani (1993) call such estimates plug-in estimates.

5.7 Examples: smooth functions of means

Here we only concern ourselves with parameters that are simple functions of expectations. These are particular examples of the general expansions of Section 4.2. As an example, consider the variance of a scalar random variable Y. Letting

$$X = \begin{pmatrix} Y \\ Y^2 \end{pmatrix},$$

then

$$\theta = \begin{pmatrix} \theta_1 \\ \theta_2 \end{pmatrix} = \begin{pmatrix} E_F[Y] \\ E_F[Y^2] \end{pmatrix}$$

and the parameter is

$$V(Y) = g(\theta) = \theta_2 - \theta_1^2.$$

The plug-in estimate of this parameter is the usual biased version of the sample variance

$$\hat{\theta} = \overline{X} = \begin{pmatrix} E_{F_n}[Y] \\ E_{F_n}[Y^2] \end{pmatrix} = \begin{pmatrix} \overline{Y} \\ \overline{Y^2} \end{pmatrix},$$

$$\hat{g} = g(\hat{\theta}) = g(\overline{X}) = \overline{Y^2} - \overline{Y}^2.$$

The expansion of this estimate may be computed from the expansions of $\overline{Y^i}$ discussed in Section 4.2.6. In most cases, the function $g(\overline{X})$ may be computed using expansions of scalar functions of single averages. For example, the estimate of the ratio of two means $\overline{Y}/\overline{Z}$ may be computed as the product of the the expansion $\overline{Y}$ times the expansion of $(\overline{Z})^{-1}$, where the expansion of the inverse is computed using Section 4.2.4.

In the following examples, cumulant expansions are computed for various familiar smooth functions. In the first of these, results are written generically in terms of $E[X^j]$. In the remaining examples, generic expressions are not easily interpreted and hence they are evaluated for particular distributions. The methods described in Chapter 2 can be used for this purpose.

The components of X, $X_1, \ldots, X_n$ are assumed to be independent and identically distributed unless otherwise stated.

5.7.1   SAMPLE MEAN AND VARIANCE

The sample mean is the simplest case possible, but it serves a useful purpose by exposing general aspects of this model. The mean is, of course, unbiased with asymptotic variance equal to the variance of a single variable y. The remaining cumulants descend in asymptotic order, reflecting a pattern that is true for any estimate.

To simplify the notation, let g represent the expansion of $\overline{Y}$ to order $O(n^{-3})$:

$$g = \mathcal{AE}_6[\overline{Y}] \to \{\widetilde{Y}, \dot{Y}, 0, 0, 0, 0, 0\}$$

where, in the following, $\mathcal{AE}_j$ denotes the asymptotic expansion to order $O(n^{-j/2})$. In the first example, the required transformation is given explicitly.

The expansions of the cumulants of g are computed by applying the transformation $\mathcal{T}_{\kappa,\mu}$ to the result computed by $\mathcal{T}_{\mu,\kappa[A]}$:

$$\mathcal{T}_{\kappa,\mu} \circ \mathcal{T}_{\mu,\kappa[A]} \circ \kappa_g \to \{\kappa_Y, 0, 0, 0, 0, 0, 0\},$$

$$\mathcal{T}_{\kappa,\mu} \circ \mathcal{T}_{\mu,\kappa[A]} \circ \kappa_{g,g} \to \left\{0, 0, \frac{\kappa_{Y,Y}}{n}, 0, 0, 0, 0\right\},$$

$$\mathcal{T}_{\kappa,\mu} \circ \mathcal{T}_{\mu,\kappa[A]} \circ \kappa_{g,g,g} \to \left\{0, 0, 0, 0, \frac{\kappa_{Y,Y,Y}}{n^2}, 0, 0\right\},$$

$$\mathcal{T}_{\kappa,\mu} \circ \mathcal{T}_{\mu,\kappa[A]} \circ \kappa_{g,g,g,g} \to \left\{0, 0, 0, 0, 0, 0, \frac{\kappa_{Y,Y,Y,Y}}{n^3}\right\},$$

from which it is clear that the kth-order cumulant of $\overline{Y}$ is of order $n^{-(k-1)}$. This holds for all cumulants of expansions. Except for the first cumulant, all expansions are for the standardized estimate $h = \sqrt{n}g$:

$$\mathcal{AE}_2[\kappa_g] \longrightarrow \kappa_Y,$$

$$\mathcal{AE}_0[\kappa_{h,h}] \longrightarrow \kappa_{Y,Y},$$

$$\mathcal{AE}_0[\kappa_{h,h,h}] \longrightarrow \kappa_{Y,Y,Y}/\sqrt{n},$$

$$\mathcal{AE}_0[\kappa_{h,h,h,h}] \longrightarrow \kappa_{Y,Y,Y,Y}/n.$$

Here the results have been written in terms of cumulants to ease interpretability and the leading term in the expansion is exact. In general, $\mathcal{AE}_0[\kappa_{h,\ldots,h}]$ evaluates to $O(n^{-(k-2)/2})$ for the kth cumulant.

The usual plug-in estimate of variance is biased where the size of the bias may be computed as

$$g = \overline{Y^2} - \overline{Y}^2,$$

$$\mathcal{AE}_2[\kappa_{h,h}] \longrightarrow \kappa_{Y,Y} - \frac{\kappa_{Y,Y}}{n}.$$

This first example serves as a test of software since we know what the results should be.

5.7.2 ONE-SAMPLE PROBLEM

In a one-sample problem where data are assumed to be normally distributed with mean 0 and unknown variance, the t-statistic with $n-1$ degrees of freedom is used. Cumulant calculations in this case show the statistic to be asymptotically standard normal and, for finite samples, the distribution of the statistic is symmetric with non-negligible kurtosis. The calculations are summarized below:

$$Law[Y] = \text{Gaussian}, \qquad \eta[Y] = \{0, \sigma^2\},$$

$$s = \sqrt{n/(n-1)(\overline{Y^2} - \overline{Y}^2)}, \quad g = \overline{Y}/s, \quad h = \sqrt{n}g,$$

$$\mathcal{AE}_2[\kappa_g] \longrightarrow 0, \qquad \mathcal{AE}_2[\kappa_{h,h}] \longrightarrow 1 + \frac{2}{n},$$

$$\mathcal{AE}_1[\kappa_{h,h,h}] \longrightarrow 0, \qquad \mathcal{AE}_2[\kappa_{h,h,h,h}] \longrightarrow \frac{6}{n}.$$

It is relatively simple to study the behaviour of the t-statistic when the assumed value of μ is incorrect, $\mu \neq 0$, by examining its cumulants:

$$Law[y] = \text{Gaussian}, \qquad \eta[y] = \left\{\mu, \sigma^2\right\},$$

$$\mathcal{AE}_2[\kappa_g] \longrightarrow \left\{\mu + \frac{5\mu}{4n}\right\}\sigma,$$

$$\mathcal{AE}_2[\kappa_{h,h}] \longrightarrow 1 + \frac{\mu^2}{2\sigma^2} + \frac{2}{n} + \frac{19\mu^2}{8\sigma^2},$$

$$\mathcal{AE}_1[\kappa_{h,h,h}] \longrightarrow \left\{\frac{5\mu^3}{4\sqrt{n}} + \frac{3\mu\sigma^2}{\sqrt{n}}\right\}\sigma^{-3},$$

$$\mathcal{AE}_0[\kappa_{h,h,h,h}] \longrightarrow \frac{\mu^4 + 3\mu^2\sigma^2 + \sigma^4}{\sigma^4 n}.$$

From these calculations, it is evident that a misspecification of μ can result in dramatic departures from the t-distribution. The test statistic is biased and skewed, with inflated variance. Asymptotically, the skewness and kurtosis go to 0, but the bias and misspecified variance remain. This simply means the statistic will fall into the rejection region more often.

5.7.3 TWO-SAMPLE PROBLEM

The two-sample problem is so similar to the previous example that we consider computing cumulants in the presence of unequal variances. If, in the construction of the t-statistic, variances, σ^2 and τ^2, are assumed equal and the sample sizes are the same, the cumulants are

$$s = \sqrt{(\overline{Y^2} - \overline{Y}^2 + \overline{X^2} - \overline{X}^2)n/(2n-2)},$$

$$g = \sqrt{2}(\overline{Y} - \overline{X})/(\sqrt{2}s), \qquad h = \sqrt{n}g,$$

$$\mathcal{AE}_2[\kappa_g] \longrightarrow 0,$$

$$\mathcal{AE}_2[\kappa_{h,h}] \longrightarrow 1 + \frac{2(\sigma^4 + \tau^4)}{n(\sigma^2 + \tau^2)^2},$$

$$\mathcal{AE}_1[\kappa_{h,h,h}] \longrightarrow 0,$$

$$\mathcal{AE}_2[\kappa_{h,h,h,h}] \longrightarrow \frac{6(\sigma^4 + \tau^4)}{n(\sigma^2 + \tau^2)^2}.$$

Asymptotically, unequal variances have no effect; however, for small sample the effect may be sizable.

5.7.4 RATIO OF MEANS

Now the parameter of interest is $g(\theta) = E[X]/E[Y]$, where $X \sim F$, $Y \sim G$. The estimate, based on two samples of equal size, is $g = \overline{X}/\overline{Y}$. Generically, expansion for the first two cumulants of $h = \sqrt{n}g$ yields

$$\mathcal{AE}_2[\kappa_g] \longrightarrow \frac{\kappa_X}{\kappa_Y} + \frac{\kappa_X \kappa_{Y,Y}}{n\kappa_Y^3} - \frac{\kappa_{X,Y}}{n\kappa_Y^2},$$

$$\mathcal{AE}_0[\kappa_{h,h}] \longrightarrow \frac{\kappa_{X,X}}{\kappa_Y^2} - \frac{2\kappa_X \kappa_{X,Y}}{\kappa_Y^3} + \frac{\kappa_X^2 \kappa_{Y,Y}}{\kappa_Y^4}.$$

Here the results have been written in terms of cumulants to ease interpretability. For example, the effect of bias would be large when $E[Y]$ is small.

5.7.5 THE CORRELATION COEFFICIENT

The correlation coefficient can be defined in terms of five means,

$$\sigma_{XY} = E[XY] - E[X]E[Y],$$

$$g(\theta) = \rho = \frac{\sigma_{XY}}{\sigma_{XX}\sigma_{YY}},$$

with estimates

$$s_{XY} = \overline{XY} - \overline{X}\,\overline{Y},$$

$$g(\hat{\theta}) = \frac{s_{XY}}{s_{XX}s_{YY}}.$$

Here cumulants can be computed directly. These rather lengthy expressions can be evaluated for any distribution. Here they are displayed for the bivariate Gaussian distribution:

$$\mathcal{AE}_2[\kappa_g] \longrightarrow \rho + \frac{\rho(\rho^2 - 1)}{2n},$$

$$\mathcal{AE}_2[\kappa_{h,h}] \longrightarrow (\rho^2 - 1)^2 + \frac{(\rho^2 - 1)^2(2 + 11\rho^2)}{2n},$$

$$\mathcal{AE}_1[\kappa_{h,h,h}] \longrightarrow \frac{\rho(\rho^2 - 1)^3}{6\sqrt{n}},$$

$$\mathcal{AE}_2[\kappa_{h,h,h,h}] \longrightarrow \frac{(\rho^2 - 1)^4(12\rho^2 - 1)}{n}.$$

5.8 Expansions of parameters

Terms of expansions of parameter estimates of Chapter 4 involved products of centred averages $\overline{X - E[X]} = E_{F_n}[X - E[X]]$. Such estimates may be considered as functionals of the empirical distribution F_n, so that $\hat{\theta} = \theta_{F_n}$. For robustness and bootstrap calculations, it is convenient to generalize the interpretation of centred averages to an arbitrary distribution G but still centred about F, so they become $E_G[X - E_F[X]]$. This generalization unifies the smooth function model of Hall (1992), and bootstrap estimation. As for exact calculations, it allows the simple computation of unbiased parameter estimates as the inverse of moment expansion.

With this generalization, the expansions of parameter estimates are considered as expansions of parameters. The only notational change required is to define a type to represent the centred expectations $E_G[\cdot] - E_F[\cdot]$ to replace the centred averages $E_{F_n}[\cdot] - E_F[\cdot]$. We denote products of averages of such terms as type $A[\nu]$.

For example, the ratio of two means in series form is

$$\theta_{F_n} = \hat{\theta} = \frac{\overline{X}}{\overline{Y}} = (\mu_X + \dot{X})(\mu_Y + \dot{Y})^{-1}$$

$$\rightarrow (\mu_X + \dot{X})\,\mu_Y^{-1}\,(1 - \mu_Y^{-1}\,\dot{Y} + (\mu_Y^{-1}\,\dot{Y})^2 + \cdots$$

$$\rightarrow \frac{\mu_X}{\mu_Y} + \left(\frac{\dot{X}}{\mu_Y} - \frac{\dot{Y}}{\mu_Y^2}\right) + \left(\frac{\mu_X \dot{Y}^2}{\mu_Y^2} - \frac{\dot{X}\dot{Y}}{\mu_Y}\right) + \cdots.$$

The expansion of the parameter is

$$\theta_G = \theta \rightarrow \frac{\mu_X}{\mu_Y} + \left(\frac{\nu_X}{\mu_Y} - \frac{\nu_Y}{\mu_Y^2}\right) + \left(\frac{\mu_X \nu_Y^2}{\mu_Y^2} - \frac{\nu_X \nu_Y}{\mu_Y}\right) + \cdots.$$

Now any moment of $\hat\theta$ is itself a parameter. When computing such moments, it will be useful to express the result in the type $A[\nu]$. All that is required is to modify the expectation operator so that each expectation $E_G[\cdot]$ is expressed as $E_F[\cdot] + E_G[(\cdot - E_F[\cdot])]$. Thus, we will write

$$E_G[X] = E_F[X] + E_G[X - E_F[X]] = \mu_X + \nu_X,$$

where $\mu_X = E_F[X]$ and $\nu_X = E_G[X - E_F[X]]$. For example, in computing the expected value of $\hat\theta$ above, the expected value of $\dot X$ becomes

$$E_G[\dot X] = E_G[(X - E_F[X])] \rightarrow 0 + \nu_X.$$

For locating the position of a product in the list representing the expansion, we assign an order to $E_G[(\cdot - E_F[\cdot])]$. Under wise usage, F will be chosen close to G, and so we treat a centred expectation like a centred average. Thus, as was the case for centred averages, the *order*, or position, of a product is determined by the number of centred expectations in the product.

   If this procedure is used to compute expansions of expectations, the expansions of moments and cumulants become expansions of parameters for an arbitrary distribution. If the distribution G is taken to be the empirical distribution F_n, the expansion is that of the plug-in estimate of the parameter. In this way, moments of estimates of moments of ... may be computed.

5.8.1   EXAMPLE: ROBUSTNESS

Suppose we wish to study the properties of an estimate under varying distributional assumptions. What is the bias and the variance, ... of a maximum likelihood estimate, defined for one distribution, when some other distribution prevails?

   We suppose that an expansion of an estimate has been computed under an assumed distribution F. This expansion involves products of averages of variables centred about their expectations *under* F. To compute the expectation under an alternative distribution, G, we may use the procedure of Section 5.2 modifying the result so that each expectation $E_G[\cdot]$ is expressed as $E_F[\cdot] + (E_G[\cdot] - E_F[\cdot])$ This expectation operator, denoted by $\mathcal{T}_{A[\nu],A}$, will be used extensively in Chapter 8.

   In the identically distributed case, objects of type $A[\nu]$ are the same as those of type ν, analogous to type μ. The transformation $\mathcal{T}_{\nu,A[\nu]} \circ \mathcal{T}_{A[\nu],A}$ is denoted by $\mathcal{T}_{\nu,A}$. Here we illustrate the application of this form of expectation for the computation of the expansion of the bias of $\hat\theta$, the maximum likelihood estimate defined in

Chapter 4:

$$\hat{\theta} \to (\theta_F) + \left(\frac{\dot{L}'}{\mu_{L''}}\right) + \left(\frac{\dot{L}'\dot{L}''}{\mu_{L''}^2} + \frac{\dot{L}'\dot{L}'\,\mu_{L^{(3)}}}{2\,\mu_{L''}^3}\right) + \cdots,$$

$$\theta \to (\theta_F) - \left(\frac{\nu_{L'}}{\mu_{L''}}\right) + \left(\frac{\nu_{L'}\nu_{L''}}{\mu_{L''}^2} - \frac{\nu_{L'}\nu_{L'}\,\mu_{L^{(3)}}}{2\,\mu_{L''}^3}\right) + \cdots,$$

$$\mathcal{T}_{\nu,A[\nu],A} \circ \hat{\theta}$$

$$\to (\theta_F) - \left(\frac{\nu_{L'}}{\mu_{L''}}\right) + \left(\frac{\nu_{L'}\nu_{L''}}{\mu_{L''}^2} + \frac{\mu_{L'}L''}{n\,\mu_{L''}^2} - \frac{\nu_{L'}\nu_{L'}\,\mu_{L^{(3)}}}{2\,\mu_{L''}^3} - \frac{\mu_{L'^2}\,\mu_{L^{(3)}}}{2\,n\,\mu_{L''}^3}\right)$$

$$+ \cdots,$$

$$\text{Bias}[\hat{\theta}] = \mathcal{T}_{\nu,A[\nu],A} \circ \hat{\theta} - \theta \to (0) + (0) + \left(\frac{\mu_{L'\,L''}}{n\,\mu_{L''}^2} - \frac{\mu_{L'^2}\,\mu_{L^{(3)}}}{2\,n\,\mu_{L''}^3}\right)$$

$$+ \left(-\frac{\mu_{L''^2}\,\nu_{L'}}{n\,\mu_{L''}^3} + \frac{3\,\mu_{L'\,L''}\,\mu_{L^{(3)}}\,\nu_{L'}}{n\,\mu_{L''}^4} - \frac{3\,\mu_{L'^2}\,\mu_{L^{(3)}}^2\,\nu_{L'}}{2\,n\,\mu_{L''}^5}\right.$$

$$- \frac{\mu_{L'\,L^{(3)}}\,\nu_{L'}}{n\,\mu_{L''}^3} + \frac{\mu_{L'^2}\,\mu_{L^{(4)}}\,\nu_{L'}}{2\,n\,\mu_{L''}^4} + \frac{\mu_{L^{(3)}}\,\nu_{L'}^2}{2\,n\,\mu_{L''}^3} - \frac{\mu_{L^{(3)}}\,\nu_{L'^2}}{2\,n\,\mu_{L''}^3}$$

$$- \frac{2\,\mu_{L'\,L''}\,\nu_{L''}}{n\,\mu_{L''}^3} + \frac{3\,\mu_{L'^2}\,\mu_{L^{(3)}}\,\nu_{L''}}{2\,n\,\mu_{L''}^4} - \frac{\nu_{L'}\,\nu_{L''}}{n\,\mu_{L''}^2} + \frac{\nu_{L'\,L''}}{n\,\mu_{L''}^2}$$

$$\left. - \frac{\mu_{L'^2}\,\nu_{L^{(3)}}}{2\,n\,\mu_{L''}^3}\right) + \cdots,$$

where terms of the same order are grouped by parentheses. The last term shows
the dependence of the bias on the expectation of the score function L' which
is non-zero, in general, when the unknown underlying distribution G is not the
hypothesized reference distribution F. Examination of the coefficients of ν suggests that this bias is likely to be smaller if the third derivatives of L are small.
This can be achieved to some degree by transforming the parameter, for example,
by working with $\log(\mu_X/\mu_Y)$ instead of the ratio μ_X/μ_Y. This calculation sup-
ports the use of transformations of the parameter to symmetrize the likelihood for
inference purposes.

5.9 Expansions of unbiased estimates

In Section 3.6.6, the derivation of an exact unbiased estimate for a product of
moments was seen to be the inverse operation of the expectation of a product of
averages. Transformations for these were given by $\mathcal{T}_{A,\mu}$ and $\mathcal{T}_{\mu,A}$, respectively.
These transformations can be interpreted as allowing the moments of one distri-
bution (F or F_n) to be written in terms of the moments of another (F_n or F). The

difference then between expectation and unbiased estimation is that the roles of F and F_n are reversed. This duality of expectation and unbiased estimation applies equally to series expansions. Once again, the roles of F and F_n are simply reversed when moving from expectation to unbiased estimation.

For example, for the ratio of two means, θ_G was expanded about F as

$$\theta_G \to \frac{\mu_X}{\mu_Y} + \left(\frac{\nu_X}{\mu_Y} - \frac{\nu_Y}{\mu_Y^2}\right) + \left(\frac{\mu_X \nu_Y^2}{\mu_Y^2} - \frac{\nu_X \nu_Y}{\mu_Y}\right) + \cdots .$$

If G is replaced with F_n, $\theta_G = \theta_{F_n} = \hat{\theta}$, and $\mathcal{T}_{\mu,A}$ applied to the expanded result, we can compute $E_F(\hat{\theta})$ to any order. However, if the roles of F and F_n are exchanged, then we have

$$\theta_F \to \frac{\overline{X}}{\overline{Y}} + \left(\frac{\nu_X}{\overline{Y}} - \frac{\nu_Y}{\overline{Y}^2}\right) + \left(\frac{\overline{X}\nu_Y^2}{\overline{Y}^2} - \frac{\nu_X \nu_Y}{\overline{Y}}\right) + \cdots ,$$

where $\nu_X = E_F(X) - E_{F_n}(X)$. Now if $\mathcal{T}_{A,\mu}$ is applied to this expansion of θ_F, an approximately unbiased estimate of θ_F results. If this is then truncated, we get the usual plug-in estimate for $E_F(\hat{\theta})$. The method may be generalized to any arbitrary parameter.

5.10   Problems

1. Compute generic expansions for the cumulants of $g(\hat{\theta})$ in the smooth function models discussed in Chapter 5.

2. Using Section 5.8, compute the variance of the maximum likelihood estimate $\hat{\theta}$.

3. Using Section 5.8, compute the mean and variance of the likelihood ratio statistic for an arbitrary distribution G close to F. Under what condition do these satisfy the condition '*variance* $\approx 2 + 2(mean - 1)$' that leads to Bartlett correction?

4. Compute cumulant expansions for the mode of the posterior density function $p(\theta|y) \propto p(\theta)l(\theta)$, where $p(\theta)$ is the prior and $l(\theta)$ is the likelihood.

5. Use the methods of this chapter to compute cumulant expansions of the signed root $\sqrt{l(\hat{\theta})}$ when θ is scalar.

6. Compute cumulant expansions of $g(\hat{\theta})$, a smooth function of the maximum likelihood estimate.

7. Compute cumulant expansions of $g(\hat{\theta})$, a smooth function of an M-estimate. How does this expansion differ from the expansion of $g(\hat{\theta})$ in the previous question?

8. Compute cumulant expansion for the Wald, $\sqrt{n}(\hat{\theta} - \theta)\hat{i}^{1/2}$, and Rao, $l(\theta)\hat{i}^{-1/2}/\sqrt{n}$, statistics.

9. Compute cumulants for the unbiased version of the sample variance assuming normality and compare these to the χ^2_{n-1} distribution. Compute them again under some other distribution. The ambitious reader may use Section 5.8 for this purpose.

10. Assuming $X_i \sim Poisson(\lambda_i)$, the Poisson dispersion test for a common rate $\lambda_i = \lambda$ has the form $(n-1)s^2/\bar{x}$. Compute the cumulants of this statistic assuming a common λ and when this assumption is false. The test statistic is asymptotically χ^2_{n-1}. Comment. Do the same for the binomial dispersion test.

6
Expansions of distributions

6.1 Introduction

Edgeworth and saddle-point methods are used to approximate the distribution of a statistic using its cumulants. Cornish–Fisher expansions are used to approximate quantiles of the distribution. The approximations are useful for application and, more importantly, for studying the properties of statistics. The approximations are elegant in the simplicity of their formulation and computation. This chapter shows how these expansions may be simply computed using the procedures of Chapters 3–5. No additional procedures are required for these calculations.

The simple derivation of Edgeworth expansions here is based on Field and Ronchetti (1990). The characteristic function, $\chi(u) = E[e^{iux}]$, of a distribution, $H(x)$, defines the cumulants, γ_j, of H:

$$\log[\chi(u)] = \sum_{j=0} \gamma_j (iu)^j / j!.$$

If G is a distribution with characteristic function $\xi(u)$ and cumulants λ_j, the Taylor expansion of $(\log[\chi(u)] - \log[\xi(u)])$ yields

$$\chi(u) = \exp\left\{\sum_{j=0}(\gamma_j - \lambda_j)(iu)^j / j!\right\} \xi(u).$$

Fourier inversion gives

$$H(x) = \exp\left\{\sum_{j=0}(\gamma_j - \lambda_j)(-D)^j / j!\right\} G(x), \tag{6.1}$$

where D denotes the differential operator. If the argument of the exponential is small, $O(n^{-1/2})$, the exponential may be approximated by its Taylor expansion computed to any order. It remains to specify the distribution G.

Many statistics defined by averages or by estimating equations have asymptotically Gaussian distributions. For this reason, we will concentrate primarily on this distribution for G. In this case, the cumulants $\lambda_j = 0$ for $j > 2$. If $\hat{\theta}$ is such a statistic with mean θ and variance σ^2/n, the standardized statistic $\hat{\delta} = n^{1/2}\sigma^{-1}(\hat{\theta} - \theta)$ has, asymptotically, a standard Gaussian distribution. In particular, its mean is 0, its variance is 1. This, together with the known order of cumulants, described below, implies that the argument of the exponential in (6.1) is $O(n^{-1/2})$, thus permitting its asymptotic expansion for $H(x)$, the distribution of $\hat{\delta}$. The cumulants γ_j of $\hat{\delta}$ may be computed from κ_j the cumulants of $\hat{\theta} - \theta$ using the relation

$$\gamma_j = \left(\frac{n}{\kappa_2}\right)^{j/2} \kappa_j. \tag{6.2}$$

6.2 The order of cumulants

The broad class of statistics described in Chapter 5 had expansions involving products of centred averages. The cumulants of such statistics have an order that depends on the number of their arguments. The greater the number of arguments, the smaller is the order of the cumulant. This order determines which cumulants enter each term of an expansion. Knowing the order of cumulants is therefore important for computing expansions.

For the statistics with expansions involving products of averages, the cumulants γ_j have known order. The kth cumulant, $k \geq 2$, has order $n^{-(k-1)}$. What follows is a proof by construction.

If $\hat{\theta}$ is defined in terms of averages or as the solution of an estimating equation, as shown in Chapter 4, its expansion involves products of averages. Because of the linearity of cumulants in their arguments ($\kappa_{a+b,c,\ldots} = \kappa_{a,c,\ldots} + \kappa_{b,c,\ldots}$) for $k > 1$ the kth cumulant is dominated by the $n^{-1/2}$ term of its expansion. Thus, the order of the kth cumulant has the same order as kth cumulant of an average namely $n^{-(k-1)}$. For a proof of another sort, see Hall (1992). The order of the cumulants of $\hat{\theta}$ and $\hat{\delta}$ are shown in Table 6.1.

In particular, since the first two cumulants of δ match those of G to order 1, the order of $\gamma_j - \lambda_j$ is at most $n^{-1/2}$ as required for the expansion of the exponential in (6.1). The order of the cumulants involved is made more explicit by expressing them in terms of coefficients of n^α:

$$\gamma_j = n^{1-j/2} \rho_j.$$

Note that there is nothing that restricts the above argument to statistics which are averages. It applies to all statistics which may be expressed in terms of products of averages.

Table 6.1. Order of cumulants, κ_j and γ_j

j	Order of κ_j	Order of γ_j
1	$n^{-1/2}$	$n^{-1/2}$
2	n^{-1}	1
3	n^{-2}	$n^{-1/2}$
4	n^{-3}	n^{-1}
5	n^{-4}	$n^{-3/2}$

6.3 Computation of Edgeworth expansions for means

The descending order of cumulants permits the expansion of the distribution (6.1), simply computed using the procedures of Chapters 4 and 5. The first step is to compute finite expansions, $\hat{\lambda}_j$ and $\hat{\gamma}_j$ of the cumulants of G and H, respectively, to some order, say $O(n^{-J/2})$. The cumulants of H are computed from the cumulants, κ_j, of $\hat{\theta}$ using (6.2). The order of higher cumulants implies that the expansion of the sum in (6.1) requires only $J + 2$ terms. The expansion of the exponential of this sum may then be computed. The specification required for computing the Edgeworth expansion is then

$$\hat{H}(x) = \exp\left\{\sum_{j=0}^{J+2}(\hat{\gamma}_j - \hat{\lambda}_j)(-D)^j/j!\right\} G(x). \tag{6.3}$$

This expansion may be simplified by noting that $\gamma_2 = 1$. Substituting the Gaussian cumulants for G: $\lambda_2 = 1$ and $\lambda_j = 0$, $j \neq 2$, and expanding the exponential in (6.3) yields an expansion whose terms in $n^{-j/2}$ may be collected. The resulting expression involves factors of the form $D^j \Phi[x]$. For some purposes, it is convenient to express these derivatives in terms of Hermite polynomials defined as

$$\mathcal{H}_j(x) = (-1)^j \phi^{(j)}(x)/\phi(x).$$

A procedure for the re-expression, $D^j \Phi[x] \rightarrow (-1)^{j-1}\mathcal{H}_{j-1}(x)\phi(x)$ is simple to define. Applying it, the expansion becomes

$$\hat{H}(x) \rightarrow \Phi(x) - \frac{\rho_3\phi(x)\mathcal{H}_2(x)}{6\sqrt{n}} - \frac{\rho_4\phi(x)\mathcal{H}_3(x)}{24n}$$

$$- \frac{\rho_3^2\phi(x)\mathcal{H}_5(x)}{72n} - \frac{\rho_5\phi(x)\mathcal{H}_4(x)}{120n^{3/2}} - \frac{\rho_3\rho_4\phi(x)\mathcal{H}_6(x)}{144n^{3/2}}$$

$$- \frac{\rho_3^2\phi(x)\mathcal{H}_8(x)}{1296n^{3/2}} - \frac{\rho_6\phi(x)\mathcal{H}_5(x)}{720n^2} - \frac{\rho_4^2\phi(x)\mathcal{H}_7(x)}{1152n^2}$$

$$- \frac{\rho_3\rho_5\phi(x)\mathcal{H}_7(x)}{720n^2} - \frac{\rho_3^2\rho_4\phi(x)\mathcal{H}_9(x)}{1728n^2} - \frac{\rho_3^4\phi(x)\mathcal{H}_{11}(x)}{31104n^2} \cdots$$

Although this form is commonly referenced, the form involving the derivative operator D, without transforming these to Hermite polynomials, is particularly useful for further computation.

For example, to compute the expansion of the density function, we need only differentiate the expansion of the distribution function. The differentiation is simply computed if the expansion involves terms of the form $D^k \Phi(x)$ since

$$D^j[D^k \Phi[x]] = D^{j+k}\Phi[X]. \tag{6.4}$$

Therefore, the Edgeworth expansion of the density may be obtained from that of the distribution by replacing each term of the form $D^k \Phi(x)$ with $D^{k+1}\Phi(x)$. The result in terms of Hermite polynomials is:

$$\hat{h}(x) \to \phi(x) + \frac{\rho_3 \phi(x)\mathcal{H}_3(x)}{6\sqrt{n}} + \frac{\rho_4 \phi(x)\mathcal{H}_4(x)}{24n}$$

$$+ \frac{\rho_3^2 \phi(x)\mathcal{H}_6(x)}{72n} + \frac{\rho_5 \phi(x)\mathcal{H}_5(x)}{120n^{3/2}} + \frac{\rho_3\rho_4 \phi(x)\mathcal{H}_7(x)}{144n^{3/2}}$$

$$+ \frac{\rho_3^3 \phi(x)\mathcal{H}_9(x)}{1296n^{3/2}} + \frac{\rho_6 \phi(x)\mathcal{H}_6(x)}{720n^2} + \frac{\rho_4^2 \phi(x)\mathcal{H}_8(x)}{1152n^2}$$

$$+ \frac{\rho_3\rho_5 \phi(x)\mathcal{H}_8(x)}{720n^2} + \frac{\rho_3^2\rho_4 \phi(x)\mathcal{H}_{10}(x)}{1728n^2} + \frac{\rho_3^4 \phi(x)\mathcal{H}_{12}(x)}{31104n^2} \cdots .$$

Approximations of higher derivatives of $H(x)$ may be computed similarly. These are useful for computing expansions of quantiles.

6.4 More general Edgeworth expansions

The above expansions, although commonly referenced, are only used for distributions of averages. Typically, the mean, θ, and variance, σ^2, of the statistic are not available. However, we can compute expansions of these cumulants and develop an Edgeworth expansion which is more generally useful.

Table 6.1 shows that the cumulants, κ_j, of $(\hat{\theta} - \theta)$ have series expansions in n^{-1} with leading terms of $n^{-(j-1)}$. Let $\kappa_{j,k}$ denote the coefficient of $n^{-k/2}$ in the expansion of κ_j. We consider the approximation of the distribution of $\hat{\delta} = n^{1/2}\kappa_{2,2}^{-1/2}(\hat{\theta}-\theta_0)$, where θ_0 is the asymptotic value of $\hat{\theta}$ and $\kappa_{2,2}/n$ is its asymptotic variance.

The series expansions of the cumulants of $\hat{\delta}$ may be found from the expansions of the cumulants of $\hat{\theta}$. If we let $\gamma_{j,k}$ denote the term of order $n^{-k/2}$ in the expansion of the jth cumulant of $\hat{\delta}$, then

$$\gamma_{j,k} = (\kappa_{2,2})^{-j/2}\kappa_{j,k+j}.$$

The Edgeworth expansion, (6.3), to order n^{-1} is therefore computed to be:

$$\hat{H}(x) \to \Phi(z) + n^{-1/2}\phi(z)\left(-\gamma_{1,2} - \frac{1}{6}\gamma_{3,4}\mathcal{H}_3(z)\right)$$

$$+ n^{-1}\phi(z)\left(-\frac{1}{2}\gamma_{1,2}^2\mathcal{H}_1(z) - \frac{1}{2}\gamma_{2,4}\mathcal{H}_1(z)\right.$$

$$\left. - \frac{1}{6}\gamma_{1,2}\gamma_{3,4}\mathcal{H}_3(z) - \frac{1}{24}\gamma_{4,6}\mathcal{H}_3(z) - \frac{1}{72}\gamma_{3,4}\mathcal{H}_3(z)\right). \tag{6.5}$$

Expressions for the $\kappa_{i,j}$ and hence the $\gamma_{i,j}$ may be computed using the procedures of Chapter 5.

We have found it useful to compute these coefficients in one step and then to substitute them in (6.5). We need, therefore, to determine how many cumulants are required and the order of the expansion of each. The Edgeworth expansion, computed to order $O(n^{-M/2})$ involves $\gamma_{j,k}$, where $k \leq M+j$, indicating the order required for the expansion of the jth cumulant. For example, in (6.5), $M=2$. The fourth cumulant, $j=4$, must be expanded to order $O(n^{-(M+j)/2}) = O(n^{-3})$ or, equivalently, $k=6$. Table 6.1 shows that the fifth cumulant is of order $O(n^{-4})$. Its expansion to $O(n^{-(M+j)/2}) = O(n^{-7/2})$ is 0. The expansions of the fifth and higher cumulants are not required. In general, an Edgeworth expansion to order $O(n^{-M/2})$ requires expansions of the cumulants $j \leq M+2$ computed to order $O(n^{-(M+j)/2})$.

6.4.1 EXAMPLE: APPROXIMATION OF THE DISTRIBUTION OF THE t-STATISTIC FOR AN ARBITRARY UNDERLYING DISTRIBUTION

In this example, we consider the distribution of the t-like statistic $\hat{\theta} = (\overline{X} - \mu_X)(\hat{\sigma}_{X,X})^{-1/2}$, where $\hat{\sigma}_{X,X}$ is the biased estimate of variance, $\overline{X^2} - (\overline{X})^2$. The statistic is just $(n-1)^{-1/2}$ times the usual t-statistic. The form used here is the same as that of the parameter estimates discussed previously.

The expansions of the cumulants of $\hat{\theta}$ may be computed and expressed in terms of cumulants. For simplicity of display, we assume, without loss of generality, that the mean, $\mu_X = 0$, and the variance, $\sigma_{X,X} = 1$. Expansions of the first four cumulants are computed to be:

$$\kappa_{\hat{\theta}} \to n^{(-1)} \frac{-\kappa_{X,X,X}}{2} + O(n^{-2}),$$

$$\kappa_{\hat{\theta},\hat{\theta}} \to n^{-1} + n^{-2}\left(3 + \frac{7\kappa_{X,X,X}^2}{4}\right) + O(n^{-3}),$$

$$\kappa_{\hat{\theta},\hat{\theta},\hat{\theta}} \to n^{-2}(-2\kappa_{X,X,X}) + O(n^{-3}),$$

$$\kappa_{\hat{\theta},\hat{\theta},\hat{\theta},\hat{\theta}} \to n^{-3}(12\kappa_{X,X,X} - 5\kappa_{X,X,X}) + O(n^{-4}).$$

It is interesting that the t-like statistic, $\hat{\theta}$, has a skewness of the sign opposite to that of X.

The asymptotic mean of $\hat{\theta}$ is 0 and its variance has leading term n^{-1}. The standardized statistic is then $\hat{\delta} = n^{1/2}\hat{\theta}$. The Edgeworth expansion is computed to be:

$$\hat{H}(z) = \Phi(z) + \phi(z)\left(\frac{1}{\sqrt{n}}\kappa_{X,X,X}\left(\frac{1}{6} + \frac{z^2}{3}\right)\right.$$

$$+ \frac{1}{n}\left(\frac{3z}{2} + \kappa_{X,X,X}^2\left(\frac{-z}{6} + \frac{z^3}{9} + \frac{z^5}{9}\right)\right.$$

$$\left.\left.+ \kappa_{X,X,X,X}\left(\frac{5z}{8} - \frac{5z^3}{24}\right)\right)\right).$$

The skewness of X affects the $O(n^{-1/2})$ term of the expansion of the tail probability. This indicates that if measurements are subject to positively skewed errors, and if this symmetry is ignored, the quoted tail probabilities will be too large in the positive tail. (Fewer *significant* results will be reported.) On the surface, this seems counter-intuitive. The amount of the increase, to order $n^{-1/2}$, at $z = 2$, the approximate upper 2.5% critical value, is $0.08\kappa_{X,X,X}n^{-1/2}$. Of course, the same thing happens in the other tail with the opposite sign. This should instill some caution in the interpretation of small t-statistics based on survival times, a situation where short survival usually attracts attention.

6.5 Cornish–Fisher expansions

Cornish–Fisher expansions are used to approximate quantiles of the distribution of a statistic. They may be simply computed from the Edgeworth expansion. We seek an expansion for z satisfying $H(z) = \alpha$. This value is the root of the equation: $H(z) - \alpha = 0$. The procedures of Chapter 4 may be used to compute the expansion of this root. These procedures required specification of the asymptotic root. Since the leading term of the Edgeworth expansion, (6.5), is $\Phi(z)$, it follows that the asymptotic root is $z_\alpha = \Phi^{-1}[\alpha]$. To compute the root, we need only specify expansions of the derivatives of $H(z)$ which, as noted above, are simple to compute if $H(z)$ is expressed in terms of $D^k\Phi(z)$.

With the derivatives defined, the procedures for computing the expansion yield:

$$\hat{H}^{-1}[\alpha] \to z_\alpha + n^{-1/2}\left(\gamma_{1,2} + \frac{1}{6}\gamma_{3,4}\mathcal{H}_2(z_\alpha)\right)$$

$$+ n^{-1}\left(\frac{1}{2}\gamma_{2,4}\mathcal{H}_1(z_\alpha) + \frac{1}{72}\gamma_{3,4}^2\mathcal{H}_1(z_\alpha)\mathcal{H}_2(z_\alpha)^2\right.$$

$$+ \frac{1}{24}\gamma_{4,6}\mathcal{H}_3(z_\alpha) + \gamma_{3,4}\mathcal{H}_2(z_\alpha)\mathcal{H}_3(z_\alpha)$$

$$\left.+ \frac{1}{72}\gamma_{3,4}^2\mathcal{H}_5(z_\alpha)\right). \tag{6.6}$$

Expansions to higher orders may be computed.

6.6 Saddle-point approximations

The terms of the above Edgeworth expansion involve Hermite polynomials of increasing order. Because the powers of the argument, z, involved are increasing, the series will converge for small z, but, may not converge for large values of $|z|$. Saddle-point approximations were developed for such large deviations. The basic idea is to *tilt* the density h so that the mean of the tilted density is close to x. The tilted density may now be approximated with a converging Edgeworth expansion since the point of expansion is close to the mean and the resulting z in (6.5) is small. The approximated density may then be tilted back to yield an approximation for h. Reid (1988) presents a very clear review for the approximations of the distributions of averages. The development here is based more closely on Field and Ronchetti (1990).

Consider the tilted density $g(x) = ce^{ax}h(x)$. The normalizing constant, c, satisfies $c^{-1} = E[e^{ax}] = e^{\mathcal{K}_H[a]}$, where $\mathcal{K}_H(t)$ is the cumulant generating function of $H(x)$:

$$\mathcal{K}_H(t) = \gamma_1 t + \gamma_2 t^2/2 + \gamma_3 t^3/3! + \cdots.$$

In this development, we have abandoned the $\chi(u)$ notation of characteristic function used by Field and Ronchetti in favour of $\mathcal{K}(t)$, in line with most discussions of saddle points.

The cumulant generating function for g is $\mathcal{K}_G(t) = \mathcal{K}_H(a+t) - \mathcal{K}_H(a)$. The mean of the density g may be computed using $\mathcal{K}'_G(0) = \mathcal{K}'_H(a)$. Equating this value to x gives the saddle-point equation:

$$\mathcal{K}'_H(a) - x = 0 \tag{6.7}$$

or, equivalently,

$$\gamma_1 + \gamma_2 a + \gamma_3 a^2/2! + \cdots - x = 0.$$

Since γ_2 is $O(1)$ and the cumulants γ_i, $i \neq 2$, have smaller order, $\leq O(n^{-1/2})$, it follows that, the asymptotic root of the equation is $a_0 = x/\gamma_{2,0}$, where $\gamma_2 = \gamma_{2,0} + n^{-1}\gamma_{2,2} + \cdots$, so that $\gamma_{2,0}$ is the asymptotic variance of $n^{1/2}(\hat{\theta} - \theta_0)$.

The order of $\mathcal{K}'_H(a_0)$ is therefore $\leq O(n^{-1/2})$. The order of the first derivative of (6.7) is 1. These are the two conditions required for application of the procedures for expanding the root of eqn (6.7) and so the expansion of the root may be computed to give the expansion of the saddle point by simply applying the procedures of Chapter 4.

The only objects which need to be specified are the expansions of the derivatives of the saddle-point equation. These are simply given by

$$\mathcal{K}_H^{\prime(i)}(a_0) = \gamma_{i+1} + \gamma_{i+2}a_0 + \gamma_{i+3}a_0^2/2! + \cdots. \tag{6.8}$$

From Table 6.1, an expansion to order $n^{-J/2}$ requires only terms up to the one involving γ_{J+2}. The expansion of the saddle point is given below to order n^{-1}:

$$\hat{a} \to x + n^{-1/2}\left(\gamma_{1,2} - \frac{x^2\gamma_{3,4}}{2}\right)$$

$$+ n^{-1}(x(\gamma_{1,2}\,\gamma_{3,4} - \gamma_{2,4}) + x^3(\gamma_{3,4}^2 - \gamma_{4,6})).$$

The terms of this expansion are dominated by terms of the form $cx(xn^{-1/2})^i$. This indicates that the series will converge for any $n > x^2$ if the cumulants are sufficiently small that they do not affect convergence. This typically happens as in common statistical practice the statistic is transformed to have an approximately Gaussian distribution, resulting in small high-order cumulants, as illustrated in the example at the end of this chapter.

With the expansion of the saddle point computed, the expansions of the cumulants of the tilted distribution, G, may be found by differentiating the cumulant generating function $\mathcal{K}_G(t) = \mathcal{K}_H(a + t)$ with respect to t and evaluating these at $t = 0$. That is, the jth cumulant is $\mathcal{K}_H^{(j)}(a)$. The expansions of the derivatives are particularly simple to compute using

$$\mathcal{K}_H^{(j)}(a) = \sum_{k=0}^{J-j+2} \gamma_{j+k} a^k / k!$$

involving cumulants of order $\leq J + 2$. The expansions of these cumulants are then used for the Edgeworth expansion of the tilted density, g, namely $\hat{g}(x + z)$ with x considered as fixed. This is an Edgeworth expansion with z as its argument. This is then evaluated at $z = 0$, corresponding to the mean of g, namely x. This involves evaluating the Hermite polynomials at $z = 0$, a rather routine task. To order n^{-1}, this expansion is computed to be

$$\hat{g}(x + 0) \to \phi(0) \left(1 + n^{-1/2} x \frac{\gamma_{3,4}}{2} + n^{-1} \left(\frac{\gamma_{1,2}\gamma_{3,4}}{2} - \frac{\gamma_{2,4}}{2} \right. \right.$$
$$\left. \left. - \frac{5\gamma_{3,4}^2}{24} + \frac{\gamma_{4,6}}{8} + x^2 \left(\frac{5\gamma_{3,4}}{8} - \frac{\gamma_{4,6}}{4} \right) \right) \right).$$

The expansion of $\log(g(x + 0))$ is then computed to be

$$\log(\hat{g}(x + 0)) \to -\frac{1}{2} \log(2\pi) + n^{-1/2} x \frac{\gamma_{3,4}}{2} + n^{-1} \left(\frac{\gamma_{1,2}\gamma_{3,4}}{2} - \frac{\gamma_{2,4}}{2} \right.$$
$$\left. - \frac{5\gamma_{3,4}^2}{24} + \frac{\gamma_{4,6}}{8} + x^2 \left(\frac{\gamma_{3,4}}{2} - \frac{\gamma_{4,6}}{4} \right) \right).$$

The complete saddle-point approximation is computed to be found by untilting the approximation by multiplying by the factor $c^{-1} \exp\{-ax\} = \exp\{K(a) - ax\}$. The expansion of the argument to order n^{-1} is:

$$K(\hat{a}) - \hat{a}x \to -\frac{x^2}{2} + n^{-1/2} \left(x\gamma_{1,2} + x^3 \frac{\gamma_{3,4}}{6} \right)$$
$$+ n^{-1} \left(-\frac{\gamma_{1,2}^2}{2} + x^2 \left(\frac{\gamma_{2,4} - \gamma_{1,2}\gamma_{3,4}}{2} \right) + x^4 \left(\frac{\gamma_{3,4}^2}{8} + \frac{\gamma_{4,6}}{8} \right) \right).$$

The resulting approximation has its logarithm given by

$$\log\left(\hat{h}(x)\right) = K(\hat{a}) - \hat{a}x + \log(\hat{g}(x+0)), \tag{6.9}$$

the components of which have been computed above. The expansion of the saddle-point approximation may thus be computed by

1. defining the expansions of $K(t)$ and its derivatives and using these for
2. computing the expansion $\hat{a}$, the root of the equation $K'(\hat{a}) = x$ and using this expansion for
3. computing expansions of the cumulants of the tilted distribution $g(x) = c\exp\{\hat{a}x\}h(x)$ and using these for
4. computing the Edgeworth expansion $\hat{g}(x+z)$ at $z = 0$ and combining the above for
5. computing the approximation of the density $\log\left(\hat{h}(x)\right) = K[a] - ax + \hat{g}(x+0)$

Each of these steps involves simple application of the procedures defined previously. The general expression (6.9) may be evaluated for any specific distribution.

6.7 Tail probabilities based on saddle points

The ultimate goal of approximating the distribution of a statistic is often the tail probability $Pr[\hat{\theta} > c]$. The approximated density may be integrated numerically. This will yield useful results if the approximation of the density is accurate over the range of integration.

Another approach is to integrate the saddle-point approximation analytically. This is possible since the approximation may be expressed in terms of a Gaussian density times a polynomial. Several authors, including Daniels (1987) and Lugannani and Rice (1980), have proposed such methods. Here, we show how the approach of Lugannani and Rice may be exploited using symbolic computation.

They showed that integration of the saddle-point density for the range $y < x$ yields the following approximation to order n^{-1}:

$$\hat{H}(x) = \Phi(\tilde{x}) + \left(\frac{1}{\tilde{x}} - \frac{1}{\tilde{z}}\right)\phi(\tilde{x}), \tag{6.10}$$

where

$$\tilde{x} = (2(\mathcal{K}_H(t) - xt))^{1/2}$$

and

$$\tilde{z} = t\left(\mathcal{K}_H^{(2)}(t)^{1/2}\right),$$

with t being the saddle point. Typically, the components of this approximation are computed numerically. However, since series expansions are available for the saddle point t, the characteristic function $\mathcal{K}$, and its derivatives, the components of (6.10) may be computed to any order. The expansion, in terms of standardized cumulants, to order n^{-1}, has the form

$$\hat{H}(x) = \Phi(A) + B\phi(A),\tag{6.11}$$

where A and B are computed as follows:

$$A \to x + n^{-1/2}\left(\gamma_{1,2} + x^2\frac{\gamma_{3,4}}{6}\right)$$

$$+ n^{-1}\left(y\left(\frac{\gamma_{1,2}\gamma_{3,4}}{3} - \frac{\gamma_{2,4}}{2}\right) + y^3\left(\frac{\gamma_{3,4}^2}{9} - \frac{\gamma_{4,6}}{24}\right)\right)$$

and

$$B \to n^{-1/2}\frac{\gamma_{3,4}}{6} - n^{-1}y\left(\frac{\gamma_{4,6}}{8} - \frac{5\gamma_{3,4}}{24}\right).$$

6.7.1 EXAMPLE: APPROXIMATIONS TO THE DISTRIBUTION OF THE RATIO OF TWO ESTIMATES OF VARIANCE

The above approximations are illustrated with the estimate of the ratio of variances, $\hat{\theta} = \hat{\sigma}_X^2/\hat{\sigma}_Y^2$, where $\hat{\sigma}_X^2 = \overline{X^2} - (\overline{X})^2$. This statistic has an $F_{n-1,n-1}$ distribution if X and Y are independent Gaussian variables with the same variance.

The expansions of the cumulants of $\hat{\theta}$ may be computed for this case by evaluating expansions of the cumulant expansions for the Gaussian distribution. This evaluation may be done in many ways. One simple way is to transform moments of the form μ_{X^i} into the corresponding Gaussian moments computed by differentiating the moment generating function $M(t) = \exp(t^2/2)$ at $t = 0$ as described in Chapter 2. The expansions of the first five cumulants are given below:

$$\kappa_{\hat{\theta}} \to 1 + 2n^{-1} + 6n^{-2} + \cdots,$$

$$\kappa_{\hat{\theta},\hat{\theta}} \to 4n^{-1} + 32n^{-2} + 204n^{-3} + \cdots,$$

$$\kappa_{\hat{\theta},\hat{\theta},\hat{\theta}} \to 48n^{-2} + 784n^{-3} + \cdots,$$

$$\kappa_{\hat{\theta},\hat{\theta},\hat{\theta},\hat{\theta}} \to 1056n^{-3} + 29280n^{-4} + \cdots,$$

$$\kappa_{\hat{\theta},\hat{\theta},\hat{\theta},\hat{\theta},\hat{\theta}} \to 34560n^{-4} + \cdots.$$

The size of third cumulant indicates that the distribution is positively skewed by a relatively large amount. For this, and many other reasons, transforming variances to log-variances is commonly recommended. Accordingly, we define $\tilde{\theta} = \log(\hat{\theta})$.

Table 6.2. Relative errors of approximations to the tail probability of $\hat{\sigma}_X^2/\hat{\sigma}_Y^2$, when X and Y are independent Gaussian with the same variance and the sample size $n = 40$

Type	$\alpha = 0.25$	$\alpha = 0.10$	$\alpha = 0.05$	$\alpha = 0.025$	$\alpha = 0.01$
Gaussian $\hat{\theta}$	0.70	0.35	0.65	1.00	1.53
Edgeworth $\hat{\theta}$	−0.014	0.0092	−0.013	−0.23	N/A
Edgeworth $\log(\hat{\theta})$	−0.035	−0.11	−0.18	−0.27	−0.42
Lugannani–Rice $\hat{\theta}$	−0.00016	0.017	0.052	0.11	0.28
Lugannani–Rice $\log(\hat{\theta})$	−0.00010	−0.00011	0.00017	0.00085	0.0027

The cumulants of the transformed statistic may be computed in exactly the same way. They result in

$$\kappa_{\tilde{\theta}} \to 0,$$

$$\kappa_{\tilde{\theta},\tilde{\theta}} \to 4n^{-1} + 8n^{-2} + \cdots,$$

$$\kappa_{\tilde{\theta},\tilde{\theta},\tilde{\theta}} \to 0,$$

$$\kappa_{\tilde{\theta},\tilde{\theta},\tilde{\theta},\tilde{\theta}} \to 32n^{-3} + \cdots,$$

$$\kappa_{\tilde{\theta},\tilde{\theta},\tilde{\theta},\tilde{\theta},\tilde{\theta}} \to 0.$$

The cumulants of the transformed statistic are much closer to their Gaussian counterparts. It will be no surprise to find that the approximations to the distribution of $\log(\hat{\theta})$ are much better than those to the distribution of $\hat{\theta}$.

The accuracy of the approximations is illustrated in Table 6.2 which presents the relative errors $(\hat{\alpha} - \alpha)/\alpha$ of the approximated tail areas $\hat{\alpha}$.

The increased accuracy associated with the logarithmic transformation is clearly evident in Table 6.2. The increased accuracy associated with the saddle-point (Lugannani and Rice) approximations is also evident.

6.8 Bibliographical notes

The material here on Edgeworth and saddle-point approximations was based largely on Field and Ronchetti (1990). Hall (1992) provides additional material. The recent work on saddle-point approximations was begun by Daniels (1954). Reid (1988) gives a particularly lucid presentation of this methodology. Easton and Ronchetti (1986) discuss generalizations as do Andrews and Feuerverger (1995). The expressions of Section 6.4 appear in Withers (1983).

6.9 Problems

1. Compute the Edgeworth expansion of the density of $\log(\overline{X^2})$ assuming X has the standard Gaussian distribution. The exact density can be found from

the χ_n density. Plot the exact density together with that of the approximations evaluated to order $1, n^{-1/2}$, and n^{-1} for $n = 20$, say. For each approximation, count the number of values for which the approximation is exact. Formulate a generalization of the Irish Clock Phenomenon (even a broken clock is correct twice a day) in terms of the order of the approximation.

2. The saddle-point approximation is exact for Gaussian and gamma distributions. This suggests that the expansion of the approximation will be accurate for χ^2 distributions. Compute the expansion of the saddle-point approximation of the density of $\log(\overline{X^2})$ assuming X has the standard Gaussian distribution. Plot this approximation, computed to $O(n^{-1})$, with the exact density.

3. The cumulants of the Poisson distribution are particularly simple. Compute the Lugannani and Rice approximation to the distribution of $\overline{X}$, where X is Poisson with mean λ. What is the exact distribution of $\overline{X}$? Compare the approximation with the exact for $n = 20$ and $\lambda = 1$. Does the discreteness of the distribution matter?

4. It is common practice when analysing data to take square roots of counts. Compute the Lugannani and Rice approximation to the distribution of $(\overline{X})^{1/2}$, where X is Poisson as above. Since the transformation is monotonic, the exact distribution is simply obtained. Compare the approximation with the exact one for the same case as above. Does the square root help?

5. Compute the Edgeworth expansion of the distribution of a general M-estimate to order n^{-1}. Investigate the conditions under which this approximation is Gaussian to order $n^{-1/2}$.

6. Compute the Edgeworth expansion of the distribution of $((n-1)/n)^{1/2}$ times the t-statistic for an arbitrary distribution and when the distribution is Gaussian. Investigate conditions under which these are the same for various orders of approximation.

7. Aspin (1948) gives a computational technique for deriving expansion for quantiles in the Behrens–Fisher problem. The technique is similar to the expansion techniques for distribution functions found in Field and Ronchetti (1990). Design an algorithm that automates the expansions for quantiles found in Aspin (1948).

7
Expansions for likelihood quantities

7.1 Introduction

Returning to the likelihood example of Section 4.5, we assume that Y is a random variable with a density or probability function f_θ, indexed by an unknown parameter θ of length p. The data $y_1, \ldots, y_n$ are assumed here to be independent realizations of Y with joint density given by the product of a contribution from each y_i:

$$f_\theta(y_1, \ldots, y_n) = \prod_{i=1}^n f_\theta(y_i).$$

Interest surrounds the parameter θ which may have some useful practical interpretation. The assumption of a parametric form for f_θ is of secondary interest and is considered restrictive though convenient.

The likelihood function

$$L(\theta) \propto f_\theta(y_1, \ldots, y_n)$$

gives a measure of 'consistency with the data' for values of θ. Typically, one considers the log-likelihood

$$l(\theta) = \log\{L(\theta)\} = \sum_{i=1}^n l_i(\theta)$$

because it is a sum and hence more convenient to study. Here $l_i(\theta) = \log\{f_\theta(y_i)\}$ is the contribution of the ith data point, y_i. The likelihood function is the basis for many inference methods. The common quantities of interest associated with $l(\theta)$ are its maximum $l(\hat\theta)$ and the deviance or distance, $w(\theta) = 2\{l(\hat\theta) - l(\theta)\}$, from this maximum. The point $\hat\theta$ is called the maximum likelihood estimate and $w(\theta)$ the Neyman–Pearson likelihood ratio statistic.

This chapter presents examples of computing expansions of these and related quantities. These show how rather complex calculations may be simply expressed using the fundamental procedures of the previous chapters. Methods for expressing terms in the summation convention notation of Section 4.6 are given. Special relations involving likelihoods are used to simplify expressions. Methods for computing Bartlett and Lawley identities are given. A simple computational proof of a central result involving Bartlett corrections is presented. And finally, procedures are described for computing expansions of the signed root of the likelihood ratio statistic. The consideration of how these results are computed leads to simple derivations and procedures.

7.2 Expansions of the estimate and deviance

In Section 4.5, the maximum likelihood estimate $\hat{\theta}$ was computed as the root of the equation

$$\overline{l_1(\hat{\theta})} = 0,$$

where l_1 denotes the vector of derivatives $l^{(1)}(y_i)$. The expansion was computed:

$$\hat{\theta} \to (\theta) + \left(-\tilde{l}_2^{-1} \cdot l_1 \right)$$
$$+ \left(\tilde{l}_2^{-1} \cdot l_2 \cdot \tilde{l}_2^{-1} \cdot l_1 - \tfrac{1}{2}\tilde{l}_2^{-1} \cdot \tilde{l}_3 \cdot \tilde{l}_2^{-1} \cdot l_1 \cdot \tilde{l}_2^{-1} \cdot l_1 \right) + \cdots,$$

where $\tilde{l}_i$ denotes the expected value of l_i and inner products are denoted by $(\cdot)$. The expansion of the average deviance $\overline{w(\theta)} = w(\theta)/n$ was computed as

$$\overline{w(\theta)} \to (0) + (0) - \left(\dot{l}_1 \cdot \tilde{l}_2^{-1} \cdot \dot{l}_1 \right)$$
$$+ \left(\dot{l}_2 \cdot \tilde{l}_2^{-1} \cdot \dot{l}_1 \cdot \tilde{l}_2^{-1} \cdot \dot{l}_1 \right.$$
$$\left. - \tfrac{1}{3}\tilde{l}_3 \cdot \tilde{l}_2^{-1} \cdot \dot{l}_1 \cdot \tilde{l}_2^{-1} \cdot \dot{l}_1 \cdot \tilde{l}_2^{-1} \cdot \dot{l}_1 \right) + \cdots,$$

where, as before, terms of like order have been grouped in parentheses.

These calculations may be generalized by replacing the likelihood function l with any function $\psi(\theta) = \psi(y, \theta)$. Such a function used in the role of a likelihood function is called a pseudo-likelihood. Although the computational procedures are the same for both likelihood and pseudo-likelihood, we describe these procedures in terms of the more familiar likelihood.

7.3 Procedures for simplification

In the summation convention of Section 4.6, a repeated index in a term of an expression identifies two arrays which are collapsed by the inner product operation. The repeated index used has no other meaning; for example, $M_{rs}v_s = M_{st}v_t$. A common representation for terms like $M_{rs}v_s$ is desired. In the context of Taylor series expansions, a common representation was accomplished before the use of indices employing integer partitions or otherwise. However, subsequent manipulation of such expressions typically requires the use of index notation. Such complex calculations can result in expressions in which the same term appears with different indices. While this does not affect the correctness of the result, a common representation is desired for its simplification. With some computation, equivalent terms may be identified. Whether or not this computation leads to more efficient calculations depends on the extent of further manipulation of the result. If there is much further work, the investment in simplification will be useful.

This section describes a few techniques to simplify expressions. They are well known, simple, perhaps even naive, but effective.

7.3.1 CANONICAL FORM

The procedure for representing a term in a common canonical form is based on two observations:

1. the term does not depend on the particular repeated indices used; and
2. all permutations of the labels of the repeated subscripts define an equivalence class of terms, each member of which is a representation of the same term.

The procedure may therefore be defined by computing all permutations of the subscripts and returning the first according to a lexicographical ordering. This will identify a unique representation of the equivalence class of terms:

1. compute R, a list of the r repeated subscripts of the term;
2. create U, a list of the form $\{U_1, \ldots, U_r\}$;
3. compute t, the term with the indices R_i replaced by U_i componentwise;
4. for each permutation $\pi[U]$ of U, compute s, the value of t with U_i replaced by $(\pi[U])_i$ and, if $s < t$ in a lexicographical ordering, set $t = s$;
5. return t.

The result may be converted to the standard indicies $\mathcal{I}$ of Section 4.6 by replacing R_i with $\mathcal{I}_i$ componentwise.

This algorithm is an elementary brute-force solution to the problem of graph isomorphism discussed in Chapter 10. The algorithm and subsequent symmetry simplifications are captured by one elegant procedure in that chapter. It has much broader implications.

7.3.2 SYMMETRY SIMPLIFICATION

Arrays involving derivatives are symmetric. For example, in the second derivative matrix, $l_{rs} = l_{sr}$. This simplification may be implemented by applying the following rule for unsorted lists:

$$l_{\{list\}} = l_{Sort[\{list\}]}$$

7.3.3 ORTHOGONAL SIMPLIFICATION

Many expressions associated with likelihood inference are invariant under linear transformations of the parameters. These expressions may be simplified if the parameter is rescaled by a matrix, so that the expected Fisher's information is equal to the identity matrix; that is, $-E[l_2] = I$. This simplification may be implemented by applying the rule

$$E[l_{rs}]A_{st\ldots}B_{t\ldots} = -A_{rs\ldots}B_{s\ldots},$$

where all indices after s in the list of indicies are decremented by 1. That is, t becomes s, u becomes t, and so on.

Applying the above simplifications to the expected value of the deviance yields

$$E[\overline{w(\theta)}] \to (0) + (0) + \left(\frac{p}{n}\right)$$

$$+ \left(\frac{I_{rrss}}{4\,n^2} + \frac{I_{rst}^2}{2\,n^2} + \frac{I_{rr\,t}\,I_{ss\,t}}{4\,n^2} - \frac{I_{stt}\,I_{r,rs}}{n^2} - \frac{I_{r,rss}}{n^2}\right.$$

$$- \frac{2\,I_{rst}\,I_{r,st}}{n^2} + \frac{I_{r,st}\,I_{s,rt}}{n^2} + \frac{I_{r,rt}\,I_{s,st}}{n^2} - \frac{I_{rs,rs}}{n^2}$$

$$\left. + \frac{I_{r,s,rs}}{n^2} + \frac{I_{r\,st}\,I_{r,s,t}}{3\,n^2}\right),$$

where, for example, $E[l_{rs}l_t]$ is denoted by $I_{rs,t}$.

7.3.4 BARTLETT IDENTITIES

The expected value of any derivative of the likelihood may be expressed in terms of lower-order derivatives. This follows from successively differentiating

$$\int f(x, \theta)\,\mathrm{d}x = \int e^{l(\theta)}\,\mathrm{d}x = 1$$

with respect to θ to get

$$E[l_r] = 0,$$

$$E[l_{rs} + l_r l_s] = 0,$$

and so on. As in Section 3.3, here the multiplicative rule for differentiation mimics the inclusion–exclusion rule and so it follows that the result after the ith iteration depends on the full partition of a set with i elements. The derivation of such identities may then be accomplished using the algorithms of Chapter 3.

7.3.5 LAWLEY IDENTITIES

In general, identities may be obtained by applying the operator $\partial_r = \partial/\partial\theta_r$ to $E[g]$:

$$\partial_r E[g] = E[\partial_r g] + E[g\,l_r].$$

For example, if $g = 1$, the Bartlett identities result. Lawley (1956) used such identities to simplify expressions involving expected likelihood derivatives by substituting $l_{stu\cdots}$ for g and solving

$$E[l_{rstu\cdots}] = \partial_r E[l_{stu\cdots}] - E[l_r l_{stu\cdots}]$$

$$I_{rstu\cdots} = \partial_r I_{stu\cdots} - I_{r,stu}.$$

Implementing these is left as an exercise.

7.4 Bartlett correction

The cumulants κ_r, $r = 1, 2, 3, \ldots$, of the deviance have a particularly interesting structure. Asymptotically, they are close to the cumulants of the χ_p^2 distribution since, as we shall see,

$$\kappa_r = 2^{r-1}(r - 1)! p + O(n^{-1}).$$

It is well known that the order of this approximation can be improved to $O(n^{-2})$ through a multiplicative correction attributed to M. S. Bartlett. The correction is based on the first cumulant which has the form

$$E[w(\theta)] = p + \frac{b}{n} + O(n^{-2}).$$

Setting B to $\{1+b/(pn)\}^{-1}$, or equivalently to $\{1-b/(pn)\}$, the Bartlett corrected deviance is $w_b(\theta) = w(\theta)B$. This correction, defined to correct the first cumulant, corrects all the cumulants to order $O(n^{-2})$.

The improvement in approximation was originally established by Lawley (1956) and again in a more elegant fashion by Barndorff-Neilsen and Cox (1979). Here we present an alternative demonstration of the result. It is simpler than Lawley's approach, in a computational sense, and more simple-minded than the approach of Barndorf-Neilsen and Cox. Our approach follows from considering the simple structure that must exist amongst the cumulants in order for the result to be true.

The approach is general, in the sense that the deviance could be based on an adjusted profile likelihood, for instance. We simply reduce the condition of Bartlett correctability down to a set of conditions which can themselves be checked by computer algebra if doing so by hand is too tedious. These conditions apply to terms in the expansion of the deviance rather than the signed root. This greatly simplifies the algebra. However, expansions for signed roots are also straightforward to automate as we shall see in the next section.

Bartlett correctability will hold if the cumulants, and hence the moments, of $w_b(\theta)$ satisfy

$$E[w_b(\theta)^r] = E\left[\left(\frac{w(\hat{\theta})}{1 + b/(pn)}\right)^r\right] = \mu_r + O(n^{-2}),$$

where μ_r is the rth moment of the χ_p^2 distribution. Equivalently, the moments of $w(\theta)$ must satisfy

$$E[w(\theta)^r] = \{1 + b/(p\,n)\}^r\,(\mu_r + O(n^{-2}))$$

$$= \mu_r + \mu_r\,r\,b/(p\,n) + O(n^{-2}). \tag{7.1}$$

In other words, the $O(n^{-1})$ term in the expansion of the rth moment of $w(\theta)$ must be $\mu_r r\,b/(p\,n)$.

If $w(\theta)$ is assumed to have the expansion

$$w(\theta) = w_0 + w_1 + w_2 + O(n^{-3/2}), \tag{7.2}$$

where subscripts identify the order of a term, then the binomial expansion

$$w(\theta)^r = w_0^r + r\,w_0^{r-1}(w_1 + w_2) + r(r-1)w_0^{r-2}w_1^2/2 + O(n^{-3/2}),$$

implies that the condition (7.1) reduces to

$$E[w_0^r + rw_0^{r-1}(w_1 + w_2) + r(r-1)w_0^{r-2}w_1^2/2]$$

$$= \mu_r + \mu_r r\,b/(p\,n) + O(n^{-2}). \tag{7.3}$$

This equation could be checked for specific values of r. However, it can be checked for *all r*. The computation required for *proving* that this equation holds for arbitrary r is illustrated below. This proof indicates how symbolic computation can be used to derive expressions where the number of the arguments, r, of the cumulant, is considered as a variable as n and p were previously.

7.4.1 LIKELIHOOD EXAMPLE

Here we consider the generic likelihood expansion of Section 4.5. To simplify the presentation, we assume $p = 1$. The proof given below is straightforward. It is not elegant. However, it demonstrates how symbolic computation can be useful for those who have neither the time nor the talent for more sophisticated approaches.

Without loss of generality, we assume $E[(l_1)^2] = -E[l_2] = 1$ as in Section 7.3.3. The expansion of the deviance divided by n in Section 4.5 to order n^{-2}, is

$$\frac{w(\theta)}{n} \to (0) + (0) + (\dot{l}_1{}^2) + \left(\dot{l}_2\dot{l}_1{}^2 + \frac{1}{3}E[l_3]\dot{l}_1{}^3\right)$$

$$+ \left(\frac{1}{3}\dot{l}_1{}^3\dot{l}_3 + \dot{l}_1{}^2\dot{l}_2{}^2 + \dot{l}_1{}^3\dot{l}_2E[l_3] + \frac{1}{4}\dot{l}_1{}^4E[l_3]^2 + \frac{1}{12}\dot{l}_1{}^4E[l_4]\right)$$

$$+ o(n^{-2}).$$

Thus, in the notation of (7.2),

$$w_0 = n\,\dot{l}_1^{\,2},$$

$$w_1 = n\,\dot{l}_1^{\,2}\left(\dot{l}_2 + \frac{1}{3}E[l_3]\dot{l}_1\right),$$

$$w_2 = n\,\dot{l}_1^{\,2}\left(\frac{1}{3}\dot{l}_1\dot{l}_3 + \dot{l}_2^{\,2} + \dot{l}_1\dot{l}_2 E[l_3] + \frac{1}{4}\dot{l}_1^{\,2}E[l_3]^2 + \frac{1}{12}\dot{l}_1^{\,2}E[l_4]\right). \qquad (7.4)$$

Hence, the condition (7.3) involves expectations with only three types of terms: $\dot{l}_1^{\,2r}$, $\dot{l}_1^{\,2r}\dot{x}$, and $\dot{l}_1^{\,2r}\dot{x}\,\dot{y}$. These are all products of $2r$ or more centred averages.

The expectation of a product of centred averages may be expressed in terms of products of cumulants. Due to Section 6.2, the term of largest order arises from the expectation of the term with the smallest number of averages, namely $\dot{l}_1^{\,2r}$. Expressing its expectation in terms of cumulants requires forming the full partition of the $2r$ averages. If the $2r$ averages are taken two at a time, they produce the term $(\kappa_{l_1,l_1})^r = n^{-r}$. All the other partitions involve cumulants with more arguments and hence are of smaller order. Only a few of these are of order $O(n^{-(r+1)})$, and the rest are smaller. The number of partitions contributing to a distinct term is simply given.

By computing examples for $r = 4,\ 5,$ and 6, the form of the general result was suggested empirically. This was then verified by a counting argument. For example, the number of partitions contributing the distinct term $(\kappa_{l_1,l_1})^{r-3}(\kappa_{l_1,l_1,l_1})^2$ is found by first ordering the $2r$ variables $\dot{l}$. This can be done in $(2r)!$ ways. The first 3 form a block of the partition that contributes the first κ_{l_1,l_1,l_1}. Since blocks are unordered sets, the same block is created in 3! ways. The second block of 3 contributes the next κ_{l_1,l_1,l_1}. Again, the same block is created in 3! ways. The next block of 2 units forms the first of the $r-3\ \kappa_{l_1,l_1}$s. Each such partition set occurs 2! times. Finally, since a partition is an unordered set of blocks, the two blocks of three lead to the same partition in 2! ways and the $r-3$ blocks lead to the same partition in $(r-3)!$ ways. The number of partitions contributing to the distinct term, $(\kappa_{l_1,l_1})^{r-3}(\kappa_{l_1,l_1,l_1})^2$, is therefore $(2r)!/(3!\,3!\,2^r\,2!\,(r-3)!)$.

These, and similar calculations show that, to order $n^{-(r+1)}$,

$$E\left[\dot{l}_1^{\,2r}\right] = C_r\left(1 + \frac{1}{n}\left(\frac{r(r-1)(r-2)}{9}\kappa_{l_1,l_1,l_1}^{\,2} + \frac{r(r-1)}{6}\kappa_{l_1,l_1,l_1,l_1}\right)\right),$$

$$E\left[\dot{l}_1 x^{2r}\dot{x}\right] = C_r\frac{1}{n}\left(r\kappa_{l_1,l_1,x} + \frac{2r(r-1)}{3}\kappa_{l_1,x}\kappa_{l_1,l_1,l_1}\right),$$

$$E\left[\dot{l}_1^{\,2r}\dot{x}\,\dot{y}\right] = C_r\frac{1}{n}\left(r\kappa_{x,y} + 2r\kappa_{l_1,x}\kappa_{l_1,y}\right),$$

where $C_r = (2n)^{-r}(2r)!/r! = n^{-r}\mu_r$.

These expectations may be simply programmed and applied to compute the expectation (7.3). The result has the required, simple dependence on r:

$$E[(w(\theta)/n)^r] \to n^{-2r}\mu_r\left(1 + r\frac{b}{n} + O(n^{-2})\right), \qquad (7.5)$$

where

$$b = -\frac{\kappa_{l_1,l_2}{}^2}{4} + \frac{\kappa_{l_2,l_2}}{4} + \frac{\kappa_{l_1,l_2}\,\kappa_{l_1,l_1,l_1}}{2} + \frac{5\,\kappa_{l_1,l_1,l_1}{}^2}{12} - \frac{\kappa_{l_1,l_1,l_2}}{2} - \frac{\kappa_{l_1,l_1,l_1,l_1}}{4}.$$

$$(7.6)$$

The form of (7.5) indicates that, when $r = 1$, the factor b is the Bartlett factor introduced above. As Lawley (1956) noted, if the moments of a statistic resemble those of a χ_p^2 distribution to order $O(n^{-2})$ so must the cumulants. The proof here is simpler: it involves the calculation of only three expectations and we have avoided the need to expand the signed root.

7.5 Signed roots of deviance

In most statistical problems, parameters are treated differently. For example, a single sample may be described in terms of a mean μ, and a variance σ^2. Usually, interest is focused on μ leading to the common t-statistic, where σ^2 is a necessary, unknown parameter but not of direct interest. (In the popular jargon, σ^2 is called a nuisance parameter.) Less frequently, the roles are reversed and the parameter of interest is σ^2. For a given sample, inference for the pair $\theta = (\mu, \sigma^2)$ may be based on a function $l(\theta)$, which we refer to as the log-likelihood, although it may represent any objective function such as a pseudo-log-likelihood.

The above situation may be generalized to cases where the parameter $\theta = \{\lambda, \psi\}$ may be partitioned into the nuisance parameters λ and those of interest, ψ. Here both are assumed to be vector valued in the development of a general algorithm. Let $\hat{\theta}_\psi = \{\hat{\lambda}_\psi, \psi\}$ be the value of θ that maximizes $l(\theta)$ with ψ fixed and let $\hat{\theta}_{\hat\psi} = \{\hat{\lambda}_{\hat\psi}, \hat\psi\}$ be the global maximum likelihood estimate for both parameters. The case of a scalar ψ is most common and so is discussed here. The approach holds for vector-valued ψ.

Inference in such situations is often based on the statistic

$$\Lambda = 2\{l(\hat{\theta}_{\hat\psi}) - l(\hat{\theta}_\psi)\},$$

whose signed root has a structure that makes its expansion easy to automate. This is illustrated in the following example.

7.5.1 EXAMPLE: GAUSSIAN MEAN AND VARIANCE

Consider the parameters of the $N(\mu, \sigma^2)$ distribution in which interest is in σ^2, the mean being considered as a nuisance parameter. In this case, $\hat{\theta}_{\sigma^2} = \{\bar{x}, \sigma^2\}$ and

$\hat{\theta}_{\hat{\sigma}^2} = \{\overline{x}, \hat{\sigma}^2\}$, where $\hat{\sigma}^2 = \overline{x^2} - (\overline{x})^2$. The average likelihood ratio, multiplied by 2 is then

$$
\begin{aligned}
\Lambda/n &= \left(\frac{\hat{\sigma}^2}{\sigma^2} - 1\right) - \log\left(\frac{\hat{\sigma}^2}{\sigma^2}\right) \\
&= \frac{\hat{\delta}}{\sigma^2} - \log\left(1 + \frac{\hat{\delta}}{\sigma^2}\right) \\
&= \hat{\delta}^2\left(\frac{1}{2\sigma^4} - \frac{\hat{\delta}}{3\sigma^6} + \frac{\hat{\delta}^2}{4\sigma^8} + \cdots\right),
\end{aligned}
$$

where $\hat{\delta} = \hat{\sigma}^2 - \sigma^2$. (Because $\hat{\delta}$ has only one non-zero component in this example, it is represented as a scalar.)

The square root of this expression is then easily found:

$$
\{\Lambda/n\}^{1/2} = \hat{\delta}\left(\frac{1}{\sqrt{2}\sigma^2} - \frac{\hat{\delta}}{3\sqrt{2}\sigma^4} + \cdots\right).
$$

In general, square roots of deviance have this form, making their expansions easy to automate.

7.5.2 NUISANCE PARAMETERS, SIGNED ROOT OF THE LIKELIHOOD RATIO

Assume both λ and ψ are vector valued. The likelihood ratio statistic has an expansion which permits the computation of its square root. This may be seen by considering the difference between the parameter estimates:

$$
\hat{\delta} = \left\{ \begin{array}{c} \hat{\delta}_\lambda \\ \hat{\delta}_\psi \end{array} \right\} = \hat{\theta}_{\hat{\psi}} - \hat{\theta}_\psi
$$

in a Taylor expansion of $l(\hat{\theta}_\psi)$ about $\hat{\theta}_{\hat{\psi}}$. Since the expansion is about the maximum, there is no linear term and so the expansion for the likelihood ratio is

$$
l(\hat{\theta}_{\hat{\psi}}) - l(\hat{\theta}_\psi) = -\sum_{i=2} l^{(i)}(\hat{\theta}_{\hat{\psi}})\,(-1)^i \frac{\hat{\delta}^{i-2}}{i!} \hat{\delta}^2 = \hat{\delta}^t\, A\, \hat{\delta},
$$

where multiplication of symbols representing arrays is by inner product. At this point in the computation, no explicit arrays are involved; only symbols. The square matrix A, defined by the above, is computed using Taylor expansions of $l^{(i)}(\hat{\theta}_{\hat{\psi}})$ about θ, once expansions of $\hat{\theta}_{\hat{\psi}}$ and $\hat{\delta}$ have been computed.

In cases where $\theta = \psi$, the expansion of the signed root is given by $\sqrt{A}\hat{\delta}$ using Section 4.2.4 to compute the expansion of the square root of the matrix $A = I + O(n^{-1})$. In all other cases, the length of ψ is less than p and the appropriate expansion is obtained by factoring $\hat{\delta} = m\,\hat{\delta}_\psi$, where, in general, m is a matrix, but a vector if ψ is scalar. As we will show, the difference in the parameter estimates, $\hat{\delta} = (\hat{\theta}_{\hat{\psi}} - \hat{\theta}_{\psi})$ factors so that we may write

$$\hat{\delta} = \left\{ \begin{matrix} \hat{\lambda}_{\hat{\psi}} \\ \hat{\psi} \end{matrix} \right\} - \left\{ \begin{matrix} \hat{\lambda}_{\psi} \\ \psi \end{matrix} \right\} = M \left\{ \begin{matrix} 0 \\ \hat{\psi} - \psi \end{matrix} \right\} = m(\hat{\delta}_{\psi}),$$

where the expansion of m is simply computed. The log-likelihood ratio may be written as

$$\hat{\delta}^t A \, \hat{\delta} = \delta_{\psi}^t m^t A m (\hat{\delta}_{\psi}),$$

and hence the square root of the log-likelihood ratio is just

$$\sqrt{m^t A m} \; (\hat{\delta}_{\psi}).$$

It remains to factor $\hat{\delta} = m(\hat{\delta}_{\psi})$. However, first we show that such a factorization is possible. We assume that $\hat{\lambda}_{\hat{\psi}}$ is a smooth function of $\hat{\psi}$ so that the Taylor expansion

$$\hat{\lambda}_{\hat{\psi}} - \hat{\lambda}_{\psi} = c_1(\hat{\psi} - \psi) + \tfrac{1}{2}c_2(\hat{\psi} - \psi)^2 + \cdots$$

$$= \left\{ c_1 + \tfrac{1}{2}c_2(\hat{\psi} - \psi) + \cdots \right\} (\hat{\psi} - \psi)$$

may be used. It follows that the difference

$$\hat{\delta} = \left\{ \begin{matrix} \hat{\lambda}_{\hat{\psi}} \\ \hat{\psi} \end{matrix} \right\} - \left\{ \begin{matrix} \hat{\lambda}_{\psi} \\ \psi \end{matrix} \right\} = \left\{ \begin{matrix} c_1 + \tfrac{1}{2}c_2(\hat{\psi} - \psi) + \cdots \\ I \end{matrix} \right\} (\hat{\delta}_{\psi})$$

$$= m(\hat{\delta}_{\psi}),$$

showing that $(\hat{\delta}_{\psi})$ is a factor of $\hat{\delta}$. The expressions for $c_j = (\partial^j/\partial\psi^j)\hat{\lambda}_{\psi}$ could be computed by implicit differentiation; however, we consider this an avoidable complication and proceed more directly.

Confident that $(\hat{\psi} - \psi)$ is a factor of $\hat{\delta}$, we compute the expansion m directly, by dividing the expansion for $\hat{\delta}$ by the expansion for $(\hat{\delta}_{\psi})$. This process of division is a simple parallel of the long division that was taught to school children before calculators were admitted into classrooms. Here, the digits representing powers of 10 are replaced by terms of descending order. (Division is only slightly more complex than multiplication.) The division algorithm is defined by

1. initialize: *divisor* $\leftarrow (\hat{\delta}_{\psi})$; *dividend* $\leftarrow (\hat{\delta})$; *quotient* $\leftarrow 0$;

2. repeat until *dividend* $= 0$,
   (a) *term* $\leftarrow$ leading term of *dividend*/ leading term of *divisor*,
   (b) *dividend* $\leftarrow$ *dividend* $-$ *term divisor*,
   (c) *quotient* $\leftarrow$ *quotient* $+$ *term*;
3. return *quotient*

This yields the expansion of m.

To accomplish the above, we need to compute an expansion for $\hat{\delta} = \hat{\theta}_{\hat{\psi}} - \hat{\theta}_{\psi}$. The expansion of the estimate $\hat{\theta}_{\hat{\psi}}$ is computed as the root of the equation $\overline{l^{(1)}(\hat{\theta}_{\hat{\psi}})} = 0$ in the ususal way using Section 4.2.5. To compute the expansion of $\hat{\theta}_{\psi}$, we note that this estimate would have resulted if the first derivative, and hence all other derivatives of the likelihood with respect to ψ, were 0. Parameters which do not affect the likelihood do not affect the estimates of other parameters. This idea is incorporated by introducing symbols for two diagonal matrices I_{λ} and I_{ψ} with 1's on the diagonal corresponding to the components of λ and ψ, respectively (and 0 elsewhere). Since $I_{\lambda} + I_{\psi} = I$, the derivatives of l may be reexpressed using $l^{(i)} \rightarrow I_{\lambda} l^{(i)} + I_{\psi} l^{(i)}$. The expansion for $\hat{\theta}_{\psi}$ is found by setting I_{ψ} to 0. The difference, $\hat{\delta}$, is then computed by subtracting the two expansions. For example, the $O(n^{-1/2})$ term of $\hat{\theta}_{\hat{\psi}}$ is $I_{\lambda}\, \dot{l}_1 + I_{\psi}\, \dot{l}_1$. The $O(n^{-1/2})$ term of $\hat{\lambda}_{\psi}$ is just $I_{\lambda}\, \dot{l}_1$, and that of $\hat{\delta}$ is $I_{\psi}\, \dot{l}_1$.

This procedure for computing expansions of roots yields the expansion for $\hat{\delta} = e_0 + e_1 + \cdots$ is

$$e_0 = 0,$$

$$e_1 = I_{\psi}\, \dot{l}_1,$$

$$\begin{aligned}
e_2 = {}& I_{\lambda}\, \dot{l}_2\, I_{\psi}\, \dot{l}_1 + I_{\psi}\, \dot{l}_2\, I_{\lambda}\, \dot{l}_1 + I_{\psi}\, \dot{l}_2\, I_{\psi}\, \dot{l}_1 \\
& + I_{\lambda}\, E[\dot{l}_3]\, I_{\lambda}\, \dot{l}_1\, I_{\psi}\, \dot{l}_1 + \tfrac{1}{2} I_{\lambda}\, E[\dot{l}_3]\, I_{\psi}\, \dot{l}_1\, I_{\psi}\, \dot{l}_1 \\
& + \tfrac{1}{2} I_{\psi}\, E[\dot{l}_3]\, I_{\lambda}\, \dot{l}_1\, I_{\lambda}\, \dot{l}_1 + I_{\psi}\, E[\dot{l}_3]\, I_{\lambda}\, \dot{l}_1\, I_{\psi}\, \dot{l}_1 \\
& + \tfrac{1}{2} I_{\psi}\, E[\dot{l}_3]\, I_{\psi}\, \dot{l}_1\, I_{\psi}\, \dot{l}_1
\end{aligned}$$

$$\vdots$$

where we assume that the parameters have been standardized so that $E[l^{(2)}(\lambda, \psi)] = -I$ as in Section 7.3.3. Further terms of smaller order may be computed by machine.

The procedure for computing the square root of the deviance is summarized by

1. compute the expansion of $\hat{\theta}_{\hat{\psi}}$;

2. compute the expansion of $\hat{\theta}_{\psi}$ by setting I_{ψ} to 0, and from it $\hat{\delta}$;

3. compute the expansion of $2\big(l(\hat{\theta}_{\hat{\psi}})/n - l(\hat{\theta}_{\psi})/n\big)$ in terms of derivatives $l^i(\hat{\theta}_{\hat{\psi}})$;

4. compute the expansion of the derivatives in terms of $l^{(i)}(\theta)$;

5. compute A, the expansion of the resulting expression after the common factor $\hat{\delta}^2$ has been factored out;

6. compute quotient m satisfying $\hat{\delta} = m(\hat{\delta}_\psi)$ by long division;

7. return $\sqrt{m^t \, A m \hat{\delta}_\psi}$.

Using this algorithm, the expansion for the average root deviance is computed to be

$$
\begin{aligned}
RLR = {}& \sqrt{2\left(\frac{l(\hat{\theta}_{\hat{\psi}})}{n} - \frac{l(\hat{\theta}_\psi)}{n}\right)} \\
\to {}& (0) + I_\psi \cdot \dot{l}_1 \\
& + I_\psi \cdot \dot{l}_2 \cdot \dot{l}_1 - \tfrac{1}{2}\dot{l}_2 \cdot I_\psi \cdot \dot{l}_1 + \tfrac{1}{2} I_\psi \cdot \tilde{l}_3 \cdot \dot{l}_1 \cdot \dot{l}_1 \\
& - \tfrac{1}{2}\tilde{l}_3 \cdot I_\psi \cdot \dot{l}_1 \cdot \dot{l}_1 + \tfrac{1}{6} I_\psi \cdot \tilde{l}_3 \cdot I_\psi \cdot \dot{l}_1 \cdot I_\psi \cdot \dot{l}_1 \\
& + \cdots .
\end{aligned}
\tag{7.7}
$$

Expansions to higher order may be computed without additional programming. To render these on paper seems pointless. However, they have been computed in a form useful for further calculation.

7.5.3 EXAMPLE: ROBUSTNESS OF THE ROOT LOG-LIKELIHOOD RATIO

The likelihood ratio is typically computed assuming an underlying distribution F. Because this distribution is typically not known precisely, it is interesting to investigate the properties of this statistic under some other distribution G. Using the procedures of Chapter 5, variance of the standardized root log-likelihood ratio (7.7) is computed to be

$$
Var_G[RLR] \to \frac{1}{n}(1 + (E_G - E_F)[(l^{(0,1)})^2 + l^{(0,2)}]) + O(n^{-2}),
$$

where $l^{(i,\,j)}$ denotes

$$
\frac{\partial^i}{\partial \lambda}\frac{\partial^j}{\partial \psi}l.
$$

Under F, $E_F[(l^{(0,1)})^2 + l^{(0,2)}] = 0$. If F is selected to be consistent with the data, F will be close to G and the expectation $(E_G - E_F)[(l^{(0,1)})^2 + l^{(0,2)}]$ may be expected to be small. If the derivatives $l^{(i)}$ are bounded, the second term will be small. Therefore, acting as though the variance is approximately $1/n$ is not unreasonable.

The mean of the statistic is rather more complicated. It is computed to be

$$
E_G[RLR] \to \frac{1}{n} \left(E_F \left[\frac{l^{(1,0)}l^{(0,2)}}{2} + \frac{l^{(0,3)}}{6} + l^{(1,0)}l^{(1,1)} + \frac{l^{(2,1)}}{2} \right] \right.
$$

$$
+ (E_G - E_F) \left[\frac{E_F[l^{(0,3)}](l^{(0,1)})^2}{6} + \frac{l^{(0,1)}l^{(0,2)}}{2} \right.
$$

$$
+ \frac{E_F[l^{(1,2)}]l^{(1,0)}l^{(0,1)}}{2} + \frac{E_F[l^{(1,2)}](l^{(1,0)})^2}{2}
$$

$$
\left. \left. + l^{(1,0)}l^{(1,1)} \right] \right) + O(n^{-2}).
$$

From this, we note that three of five terms in the argument of $E_G - E_F$ involve the expectation under F of the third derivative of l. If the log-likelihood has expected third derivative equal to 0, the unknown component of the bias, the term in $E_G - E_F$, will have fewer terms and may be expected to be much smaller. For this reason, choosing parameters defined by log-likelihoods with small expected third derivative will tend to yield inference statements less sensitive to the particular assumed distribution. The inferences will be more robust.

7.6 Bibliographical notes

In this chapter, many standard asymptotic calculations for likelihood quantities were computed. We have restricted our attention to just the technical details of the particular calculations and offered very little in the way of theoretical, practical or geometric insight into the quantities derived. An excellent treatment of such issues can be found in McCullagh (1987) and Barndorff-Nielsen and Cox (1989, 1994). The issue of bias and variance corrections of the maximum likelihood estimate has been studied extensively by DiCiccio (1984) and Davison and Stafford (1996). Firth (1993) takes an interesting approach to the problem of bias by suggesting a correction of the score rather than directly correcting the maximum likelihood estimate itself. This approach is similar to corrections of the profile score function although motivated by a different problem. Barndorff-Nielsen (1983), Cox and Reid (1987) and McCullagh and Tibshirani (1990) present various adjustments of the profile likelihood arguments to correct for the effect of nuisance parameter estimation. All of these have the effect of reducing the bias of the profile score function by an order. DiCiccio and Stern (1994) have shown the Bartlett correctabilty of these adjusted likelihoods. Stafford (1996) suggested an observed variance correction similar to that of McCullagh and Tibshirani (1990) to adjust for the effect of a model misspecification.

The long division algorithm of Section 7.5 is a machine implementation of a standard hand technique taught by Miss Cullen in 1952 in grade 4. It is a common hand-calculation technique used, for example, by DiCiccio, Hall, and

Romano (1991), although this is not mentioned in the paper. Barndorff-Nielsen and Blaesild (1986) give a computer algebra procedure for the evaluation of the Bartlett correction for the usual likelihood ratio statistic. Stafford, Andrews, and Wang (1993) give the first symbolic algorithms for likelihood inference.

7.7 Problems

1. Devise a test of the expansion algorithms by computing terms in the expansion of $l^{(1)}(\hat{\theta})$ in both the scalar and vector cases.

2. Show the Bartlett correction of Section 7.4 is equivalent to that of McCullagh (1987).

3. Given the adjustment to the score function suggested by Firth (1993), determine its effect on the Bartlett correction by direct generic computation in the scalar case. Evaluate it for the Normal and Gamma examples of Chapter 2.

4. Assuming ψ and λ are both scalar quantities, compute generically the Bartlett correction of the adjusted profile likelihood of Cox and Reid (1987), and McCullagh and Tibshirani (1990).

5. Compute the variance and skewness of the maximum likelihood estimate in the scalar and vector cases.

6. Now differentiation satisfies the same inclusion–exclusion rules as full partitions as noted in Chapter 3. It follows that if the expected value of any derivative $E[l^{(k)}]$ is indexed by a list of k subscripts, the full partition of this list may be computed and each component associated with an expected value of a derivative. Equating this expression to 0 gives one of the Bartlett identities. For example, the full partition of the list $\{r, s\}$ is $\{\{r\}, \{s\}\} + \{\{r, s\}\}$, which can be mapped to $E[l_r l_s] + E[l_{rs}]$. The Bartlett identities may be implemented by applying the rule

$$E[l^{(k)}_{\{list\}}] \rightarrow \{\{list\}\} - \mathcal{P}(\{list\})$$

and considering the result to be of type μ from Chapter 3. For example, this simplification yields the computed result:

$$I_{rst} \rightarrow -I_{r,s,t} - I_{r,st} - I_{s,rt} - I_{t,rs}.$$

8
The analytic bootstrap

8.1 Introduction

Bootstrap methodology refers to a class of general resampling procedures for inference. The book by Efron and Tibshirani (1993) is an excellent introduction to the subject. Many of the procedures they describe involve the use of averages of a large number of Monte Carlo samples to approximate an expectation. In this chapter, the procedures of Chapter 5 are used to compute series expansions of the same expectations. The evaluation of these series expansions is much faster than Monte Carlo sampling because it involves only the evaluation of a small number of averages. In addition, the analytic form of the expansions permits further analysis. For example, the moments and cumulants of the bootstrap estimates may be computed.

These calculations are illustrated with a number of simple parameters. Bootstrap estimates of variance, confidence intervals, and BCA confidence intervals for these parameters are computed. The same, fundamental procedures used for these examples may be applied to much more complex parameters.

The computational procedures are beautiful in their simplicity. Appreciating the simplicity is difficult. Using the methods of Section 5.8, parameters of a distribution G are expanded about another, reference distribution F. When G is replaced with the empirical distribution F_n, we immediately have the plug-in estimate in expanded form appropriate for further series computation. Alternatively, it may be evaluated to give the usual plug-in estimate. This is a good strategy because it simplifies computation although admittedly it is novel.

We make heavy use of the fact that the empirical distribution, F_n, is a distribution like any other. Procedures for computing expectations can be used to compute expressions involving expectations with respect to F_n. But such expectations are just averages, and expressions evaluated for F_n are the bootstrap plug-in estimates, the quantities commonly estimated by bootstrap Monte Carlo. We will refer to these quantities as bootstrap estimates. These ideas are illustrated below with simple examples which, in some cases, lead to complex expressions.

8.2 Variance

This section presents the symbolic computation of the bootstrap estimate of variance, the bootstrap estimate of the variance of this estimate, and the variance of *this* estimate, thus demonstrating the effectiveness of retaining a series form for all plug-in estimates.

8.2.1 SERIES IN, SERIES OUT

It is convenient if all objects have the same series form for a sequence of calculations. These computations require the use of only one procedure from Chapter 5. The simplicity of the computations follows from the expansion of parameters discussed in Section 5.8.

The variance of G, the distribution of a random variable, X, is defined by

$$\sigma^2 = E_G[X^2] - (E_G[X])^2. \tag{8.1}$$

In Section 5.8, expectations with respect to the distribution G were expanded about the distribution F by writing

$$E_G[Y] = E_F[Y] + (E_G[Y] - E_F[Y])$$
$$= \mu_Y + \nu_Y,$$

where μ_Y denotes $E_F[Y]$ and ν_Y denotes $E_G[Y] - E_F[Y]$ for any Y. With this notation, (8.1) becomes

$$\sigma^2 = \mu_{X^2} + \nu_{X^2} - (\mu_X + \nu_X)^2$$
$$= (\mu_{X^2} - \mu_X^2) + (\nu_{X^2} - 2\mu_X\nu_X) - (\nu_X^2), \tag{8.2}$$

where, in the last line, terms are grouped in powers of ν. This object has the same form as the expansion of parameter estimates.

The series form for the plug-in estimate of σ^2 is found by replacing the distribution G with the empirical distribution F_n in (8.2). Expectations, $\nu_Y = E_G[Y] - E_F[Y]$, then become centred averages, previously denoted by $\dot{Y}$. The estimate has the form

$$\widehat{\sigma^2} = \left(\mu_{X^2} - (\mu_X)^2\right) + \left(\dot{X^2} - 2\mu_X\dot{X}\right) - \left(\dot{X}^2\right). \tag{8.3}$$

This plug-in estimate has the same computer representation (8.2) as the parameter. The only change is in the way this object is interpreted. The use of centred expectations ν for the representation of parameters has eliminated the computation for producing plug-in estimates. In this sense, expansions for bootstrap estimates are produced for free, without *any* further computation.

The usual bootstrap estimate can be evaluated by explicitly writing each $\dot{Y}$ as $\overline{Y} - E_F[Y]$. However, this required computation is inefficient. Just as the expansion of $l(\theta)$ is simpler if the expansion is about $\hat{\theta}$, here simpler expansions result from expanding about $F = F_n$. In this case, all of the averages of centred variables are

$$\dot{Y} = E_{F_n}[Y] - E_F[Y] = 0$$

Table 8.1. Scores of 30 students in two mathematics achievement tests (from Snedecor and Cochrane, 1967).

I	73	41	83	71	39	60	51	41	85	88
II	29	24	34	27	24	26	35	18	33	39
I	44	71	52	74	50	43	85	53	85	44
II	27	35	25	29	13	13	40	23	40	22
I	66	60	33	43	76	51	57	35	40	76
II	25	21	26	19	29	25	19	17	17	35

and all expectations, when evaluated, have the form

$$\mu_{X^a} = E_{F_n}[X^a] = \overline{X^a}.$$

This gives the usual bootstrap plug-in estimate.

Snedecor and Cochrane (1967, p. 180) give an example (see Table 8.1) of two mathematics scores of 30 students.

The estimate of variance is applied to the first mathematics score of each of the 30 students. The resulting estimate of variance, $\widehat{\sigma^2}$, is 290.133. Note that this estimate is the sample variance multiplied by a factor 29/30. The computation of plug-in (bootstrap) estimates of much more complex parameters is just as simple.

8.2.2   VARIANCE OF THE ESTIMATE

The next challenge is to estimate the standard error of $\widehat{\sigma^2}$. If the data were assumed to be Gaussian, the variance of the $\widehat{\sigma^2}$ would be found, from the χ^2 distribution, to be

$$2(n-1)n^{-2}\sigma^4, \tag{8.4}$$

with estimate

$$2(n-1)n^{-2}(\widehat{\sigma^2})^2.$$

However, if the data are not Gaussian, the expression for the variance is typically more complicated. Fortunately, it can be computed using the procedures of Chapter 5.

The series expansion of the *variance* of $\widehat{\sigma^2}$ is computed by applying the transformation $\mathcal{T}_{v,\kappa[A]}$, defined in Section 5.8.1, to the list $\{\sigma^2, \widehat{\sigma^2}\}$ representing the second cumulant of the statistic. The series is expressed in terms of centred expectations ν:

$$\mathcal{T}_{\nu,\kappa[A]} \circ \{\widehat{\sigma^2}, \widehat{\sigma^2}\}$$

$$\to (0) + (0)$$

$$+ \left(\frac{-4\mu_X^4}{n} + \frac{8\mu_X^2 \mu_{X^2}}{n} - \frac{\mu_{X^2}^2}{n} - \frac{4\mu_X \mu_{X^3}}{n} + \frac{\mu_{X^4}}{n} \right)$$

$$+ \left(\frac{-16\mu_X^3 \nu_X}{n} + \frac{16\mu_X \mu_{X^2} \nu_X}{n} - \frac{4\mu_{X^3} \nu_X}{n} + \frac{8\mu_X^2 \nu_{X^2}}{n} \right.$$

$$\left. - \frac{2\mu_{X^2} \nu_{X^2}}{n} - \frac{4\mu_X \nu_{X^3}}{n} + \frac{\nu_{X^4}}{n} \right)$$

$$+ \left(\frac{10\mu_X^4}{n^2} - \frac{20\mu_X^2 \mu_{X^2}}{n^2} + \frac{4\mu_{X^2}^2}{n^2} + \frac{8\mu_X \mu_{X^3}}{n^2} - \frac{2\mu_{X^4}}{n^2} \right.$$

$$\left. - \frac{24\mu_X^2 \nu_X^2}{n} + \frac{8\mu_{X^2} \nu_X^2}{n} + \frac{16\mu_X \nu_X \nu_{X^2}}{n} - \frac{\nu_{X^2}^2}{n} - \frac{4\nu_X \nu_{X^3}}{n} \right)$$

$$+ \left(\frac{40\mu_X^3 \nu_X}{n^2} - \frac{40\mu_X \mu_{X^2} \nu_X}{n^2} + \frac{8\mu_{X^3} \nu_X}{n^2} - \frac{16\mu_X \nu_X^3}{n} \right.$$

$$\left. - \frac{20\mu_X^2 \nu_{X^2}}{n^2} + \frac{8\mu_{X^2} \nu_{X^2}}{n^2} + \frac{8\nu_X^2 \nu_{X^2}}{n} + \frac{8\mu_X \nu_{X^3}}{n^2} - \frac{2\nu_{X^4}}{n^2} \right)$$

$$+ \left(\frac{-6\mu_X^4}{n^3} + \frac{12\mu_X^2 \mu_{X^2}}{n^3} - \frac{3\mu_{X^2}^2}{n^3} - \frac{4\mu_X \mu_{X^3}}{n^3} + \frac{\mu_{X^4}}{n^3} \right.$$

$$+ \frac{60\mu_X^2 \nu_X^2}{n^2} - \frac{20\mu_{X^2} \nu_X^2}{n^2} - \frac{4\nu_X^4}{n} - \frac{40\mu_X \nu_X \nu_{X^2}}{n^2}$$

$$\left. + \frac{4\nu_{X^2}^2}{n^2} + \frac{8\nu_X \nu_{X^3}}{n^2} \right) + \left(\frac{-24\mu_X^3 \nu_X}{n^3} + \frac{24\mu_X \mu_{X^2} \nu_X}{n^3} \right.$$

$$- \frac{4\mu_{X^3} \nu_X}{n^3} + \frac{40\mu_X \nu_X^3}{n^2} + \frac{12\mu_X^2 \nu_{X^2}}{n^3} - \frac{6\mu_{X^2} \nu_{X^2}}{n^3}$$

$$\left. - \frac{20\nu_X^2 \nu_{X^2}}{n^2} - \frac{4\mu_X \nu_{X^3}}{n^3} + \frac{\nu_{X^4}}{n^3} \right)$$

$$+ \left(\frac{-36\mu_X^2 \nu_X^2}{n^3} + \frac{12\mu_{X^2} \nu_X^2}{n^3} + \frac{10\nu_X^4}{n^2} + \frac{24\mu_X \nu_X \nu_{X^2}}{n^3} \right.$$

$$- \frac{3\nu_{X^2}^2}{n^3} - \frac{4\nu_X \nu_{X^3}}{n^3} ,$$

$$\left. \frac{-24\mu_X \nu_X^3}{n^3} + \frac{12\nu_X^2 \nu_{X^2}}{n^3} , \frac{-6\nu_X^4}{n^3} \right).$$

$$\tag{8.5}$$

It could be evaluated for any specific distribution G and the result compared to the Gaussian approximation. As in Section 5.8.1, terms in v represent the contribution to the variance under model misspecification.

8.2.3 THE BOOTSTRAP ESTIMATE OF THE VARIANCE OF THE ESTIMATE

Rather than assuming a Gaussian or other specific distribution for the data, we may *compute* $\widehat{\widehat{\sigma^2}}$, the bootstrap estimate of (8.5), where the result is in series form. This bootstrap estimate has the same computer representation as (8.5). As for (8.2) and (8.3), the *computation* that takes place is just in the interpretation of this computer representation.

As in the case of the statistic $\widehat{\sigma^2}$, the resulting series expansion may be *evaluated* using any F, the choice of $F = F_n$ being particularly convenient. For the mathematics scores of 30 students given by Snedecor and Cochrane (1967) the variance of $\widehat{\sigma^2}$ is 1971.26. This same answer would be obtained by an infinite standard bootstrap simulation. The analytic calculation is exact and fast. The estimate is very much less than that suggested by (8.4) because the distribution of the scores has much shorter tails than suggested by the Gaussian distribution.

The value of the series expansion is dominated by its initial terms. This can be seen by evaluating the terms of (8.5). This gives the series $\{0, 0, 1916.04, 0, 59.32, 0, -4.11\}$. The leading non-zero term contributes 97% of the result. In subsequent calculations, to avoid needless precision, we will work with a truncated version of (8.5).

8.2.4 VARIANCE OF THIS BOOTSTRAP ESTIMATE

Now, how good is this bootstrap estimate? What is the variance of the bootstrap estimate? Since $\widehat{\widehat{\sigma^2}}$ is in series form, its variance may be computed in the same way that the variance of $\widehat{\sigma^2}$ was found, namely by applying the transformation $\mathcal{T}_{v,\kappa[A]}$ to the list $\{\widehat{\widehat{\sigma^2}}, \widehat{\widehat{\sigma^2}}\}$. To simplify the result we consider only the leading term:

$$
\mathcal{T}_{v,\kappa[A]}\left[\left\{\widehat{\widehat{\sigma^2}}, \widehat{\widehat{\sigma^2}}\right\}\right] \to n^{-3}\big(-256\mu_X^8 + 1024\mu_X^6\mu_{X^2}
$$
$$
-1152\mu_X^4\mu_{X^2}^2 + 352\mu_X^2\mu_{X^2}^3 - 4\mu_{X^2}^4 - 512\mu_X^5\mu_{X^3}
$$
$$
+832\mu_X^3\mu_{X^2}\mu_{X^3} - 224\mu_X\mu_{X^2}^2\mu_{X^3} - 128\mu_X^2\mu_{X^3}^2
$$
$$
+32\mu_{X^2}\mu_{X^3}^2 + 224\mu_X^4\mu_{X^4} - 208\mu_X^2\mu_{X^2}\mu_{X^4}
$$
$$
+8\mu_{X^2}^2\mu_{X^4} + 48\mu_X\mu_{X^3}\mu_{X^4} - \mu_{X^4}^2 - 96\mu_X^3\mu_{X^5}
$$
$$
+48\mu_X\mu_{X^2}\mu_{X^5} - 8\mu_{X^3}\mu_{X^5} + 32\mu_X^2\mu_{X^6}
$$
$$
-4\mu_{X^2}\mu_{X^6} - 8\mu_X\mu_{X^7} + \mu_{X^8}\big). \tag{8.6}
$$

To better understand the precision of the bootstrap estimate $\widehat{\sigma^2}$, we compute the coefficient of variation assuming that F is a standard Gaussian distribution. To compute this coefficient, (8.6) and (8.5) were evaluated by substituting μ_{X^a} with the corresponding Gaussian moments. The resulting coefficient of variation has leading term $14n^{-1}$. The root of this coefficient, when $n = 30$, is approximately 0.7, indicating that the bootstrap estimate of variance of $\widehat{\sigma^2}$ has a standard error of 70% of its estimate—*if* the data were Gaussian. For other distributions, it may be much higher.

8.2.5 APPROXIMATE CONFIDENCE LIMITS

It is standard practice in statistics to set aside concerns about the precision of estimates of standard error and use them to compute approximate confidence limits. Approximate confidence intervals can be calculated with the form $\hat{\theta} \pm 1.96\widehat{\sigma}_{\hat{\theta}}$, where $\widehat{\sigma}_{\hat{\theta}}$ is the square root of the bootstrap estimate of variance of the statistic. For the variance of Snedecor and Cochrane mathematics scores, the approximate confidence interval is $\{204, 376\}$. The use of this symmetric interval is grounded in the general robustness of the t-statistic.

8.3 BCA confidence limits

The standard interval for variance based on the χ^2 distribution, which arises from assuming that the data are Gaussian, is not symmetric. This asymmetry arises from the asymmetry of the distribution of the estimate. This asymmetry, as we shall see below, is typically associated with a relation between the value of a parameter and the variance of its estimate.

BCA (bias-corrected accelerated) confidence intervals, described by Efron and Tibshirani (1993), are based on the assumption that, in many cases, the variance of the estimate of a parameter or, equivalently, its standard error depends locally on the parameter in a linear fashion:

$$\sigma_{\hat{\theta}} \approx a + b\theta. \tag{8.7}$$

The use of $\sigma_{\hat{\theta}}$ rather than $\sigma_{\hat{\theta}}^2$ above leads to a slope b which is scale invariant and hence, as is typical for such parameters, more precisely estimated.

We imagine a large collection of distributions close to F and the pair $\{\theta, \sigma_{\hat{\theta}}\}$ associated with each distribution. The slope b could be estimated by computing the straight line regression of $\sigma_{\hat{\theta}}$ on θ. Now, continuing with our thought Monte Carlo, we imagine generating a distribution close to F by drawing a random sample from F and using the resulting empirical distribution F_n. Repeating this would produce, if we did it, a large number of pairs $\{\hat{\theta}, \widehat{\sigma}_{\hat{\theta}}\}$, one for each distribution F_n. The regression would now be a regression involving estimates, of $\widehat{\sigma}_{\hat{\theta}}$ on $\hat{\theta}$. But

$\widehat{\sigma}_{\hat{\theta}} = \sqrt{\widehat{\sigma_{\hat{\theta}}^2}}$ has the same series form as that of $\hat{\theta}$.

From (8.7), the coefficient b above may be estimated by regressing bootstrap estimates of standard errors on the estimates themselves. The Gauss Markov estimate of the slope of a straight line regression of Y on X is just $Cov[X, Y]/Var[X]$.

Here, the slope of the regression line may be estimated symbolically using bootstrap estimates of the variance and covariance involved:

$$\hat{b} = \widehat{Cov[\hat{\sigma}_{\hat{\theta}}, \hat{\theta}]}/\widehat{\sigma^2_{\hat{\theta}}}.$$

Just as the bootstrap estimate of the variance of a statistic can be computed analytically, so can the bootstrap estimate of the covariance of two statistics. Since both the square root of the analytic bootstrap estimate of variance, $\hat{\sigma}_{\hat{\theta}}$, and the statistic $\hat{\theta}$ are represented as series, the covariance of $\hat{\sigma}_{\hat{\theta}}$ and $\hat{\theta}$

$$Cov[\hat{\sigma}_{\hat{\theta}}, \hat{\theta}] = E_G[\hat{\sigma}_{\hat{\theta}}\hat{\theta}] - E_G[\hat{\sigma}_{\hat{\theta}}]E_G[\hat{\theta}]$$

can be calculated as $\mathcal{T}_{v,\kappa[A]} \circ \{\hat{\sigma}_{\hat{\theta}}, \hat{\theta}\}$. The bootstrap estimate of this covariance is computed by evaluating the expectation at $G = F_n$ described above.

This estimated slope $\hat{b}$ is used to make approximate confidence limits following the steps:

$$\text{(i)} \quad \theta - \hat{\theta} = c\sigma_{\hat{\theta}},$$

$$\text{(ii)} \quad \sigma_{\hat{\theta}} \approx \hat{\sigma}_{\hat{\theta}} + (\theta - \hat{\theta})\hat{b}, \tag{8.8}$$

$$\text{(iii)} \quad \theta = \hat{\theta} + c\hat{\sigma}_{\hat{\theta}}(1 - c\hat{b})^{-1}.$$

The limit (8.8) can be evaluated using the bootstrap estimates $\hat{\sigma}_{\hat{\theta}}$ and $\hat{b}$. The resulting interval for the 30 mathematics scores is $\{209, 381\}$.

The BCA limits are represented as series involving products of averages. Their statistical properties can be studied as in Section 8.2. For example, the expected length of the BCA interval may be calculated by applying the expectation operator to the difference of the expansions of the upper and lower limits. It is clear that some preliminary algebraic simplification is possible but such hand calculation may not make the best use of the statisticians time.

8.4 Correlation

The simple calculations applied above to the estimate of variance may be applied to much more complicated statistics. Here we consider the estimation of correlation.

The correlation of two variables X and Y, $\rho_{X,Y}$, may be defined in terms of cumulants:

$$\rho_{X,Y} = \kappa_{X,Y}(\kappa_{X,X}\kappa_{Y,Y})^{-1/2}, \tag{8.9}$$

where each covariance $\kappa_{X,Y}$ is defined in terms of expectations:

$$\kappa_{X,Y} = E_G[XY] - E_G[X]E_G[Y]. \tag{8.10}$$

If expectations E_G are expressed as

$$E_G[X] = \mu_X + \nu_X,$$

the procedures of Chapter 5 may be used with (8.5) and (8.10) to give a series expansion of $\rho_{X,Y}$.

8.4.1 ESTIMATE OF CORRELATION

The bootstrap estimate of correlation may be computed in the manner of Section 8.2. This leads to the following series form of the estimated correlation coefficient, shown here to order n^{-1}:

$$
r_{X,Y} = (\rho_{XY}) + \left(\dot{X}\dot{Y} - \frac{\dot{X}^2 \rho_{XY}}{2} - \frac{\dot{Y}^2 \rho_{XY}}{2} \right)
$$

$$
+ \left(-(\dot{X}\dot{Y}) - \frac{\dot{X}^2(\dot{X}\dot{Y})}{2} - \frac{(\dot{X}\dot{Y})\dot{Y}^2}{2} + \frac{\dot{X}^2 \rho_{XY}}{2} \right.
$$

$$
\left. + \frac{3\dot{X}^{2^2} \rho_{XY}}{8} + \frac{\dot{Y}^2 \rho_{XY}}{2} + \frac{\dot{X}^2 \dot{Y}^2 \rho_{XY}}{4} + \frac{3\dot{Y}^{2^2} \rho_{XY}}{8} \right) + O(n^{-3/2}).
$$

This series may be evaluated for any empirical distribution in the same way that the estimate of variance was evaluated above. The value for the 30 pairs of mathematics scores of Snedecor and Cochrane (1967) is 0.7725.

8.4.2 THE VARIANCE OF THE ESTIMATE

The series expansion of the variance of $r_{X,Y}$ may be computed by applying the transformation $\mathcal{T}_{\mu,\kappa[A]}$ to the list $\{r_{X,Y}, r_{X,Y}\}$ representing the second cumulant of $r_{X,Y}$:

$$
\mathcal{T}_{\mu,\kappa[A]}[\{r_{X,Y}, r_{X,Y}\}] \to (0) + (0)
$$

$$
+ \left(-\frac{\mu_{XY}^2}{n} + \frac{\mu_{X^2Y^2}}{n} + \frac{\mu_{X^2}\mu_{XY}\rho_{XY}}{n} - \frac{\mu_{X^3Y}\rho_{XY}}{n} \right.
$$

$$
+ \frac{\mu_{XY}\mu_{Y^2}\rho_{XY}}{n} - \frac{\mu_{XY^3}\rho_{XY}}{n} - \frac{\mu_{X^2}^2 \rho_{XY}^2}{4n} + \frac{\mu_{X^4}\rho_{XY}^2}{4n}
$$

$$
\left. - \frac{\mu_{X^2}\mu_{Y^2}\rho_{XY}^2}{2n} - \frac{\mu_{Y^2}^2 \rho_{XY}^2}{4n} + \frac{\mu_{X^2Y^2}\rho_{XY}^2}{2n} + \frac{\mu_{Y^4}\rho_{XY}^2}{4n} \right)
$$

$$
+ O(n^{-2}),
$$

where terms are grouped by their order. The bootstrap estimate of this variance is found by evaluating this series with $F = F_n$ by replacing each expectation μ with the corresponding average and ρ_{XY} with the correlation of the data. The value of this bootstrap estimate of variance for the Snedecor and Cochrane mathematics scores is 0.004782.

The expression may be evaluated for F, the standard Gaussian distribution by replacing each expectation μ with the corresponding Gaussian moment found by differentiating the moment generating function of the standard bivariate Gaussian distribution and setting $\rho_{XY} = \rho$. The value of the leading term of the series representing this bootstrap estimate for Gaussian data is $(1 - \rho^2)^2/n$.

8.4.3 APPROXIMATE CONFIDENCE LIMITS

The bootstrap estimate of variance may be used to compute approximate confidence limits. These are $\{0.64, 0.91\}$.

8.4.4 BCA CONFIDENCE LIMITS

This symmetric interval does not take into account the bounded nature of the correlation parameter. If the data are Gaussian, the dependence of the variance of the estimate, namely $(1 - \rho^2)^2/n$, on the correlation ρ may be used to define Fisher's variance stabilizing transformation $t = 1/2 \log[(1 + \rho)/(1 - \rho)]$.

The series expansion of t may be computed using the procedures of Chapter 4. The variance of t is computed to be approximately $n^{-1} + O(n^{-2})$ if the data are Gaussian. For other cases, the variance may be estimated. The bootstrap estimate of the variance of t may be calculated and used to compute an approximate confidence interval for t. The confidence limits based on t can be back-transformed to limits for ρ. The limits resulting from application of this approach to the two mathematics scores are $\{0.60, 0.88\}$. This interval is appropriately asymmetric.

While this transformation is appropriate for Gaussian data it is not clear that it is useful for the unknown distribution at hand. The BCA intervals are not based on a Gaussian assumption. The BCA limits may simply be computed by applying the procedures of Section 8.3 to r_{XY}. For the two mathematics scores, the limits are $\{0.57, 0.84\}$, similar in their asymmetry to those based on the transformed correlation. In this case, however, we are not concerned about the appropriateness of a transformation based on the specific Gaussian distribution.

8.5 Theoretical basis for BCA intervals

The computational procedures may be used to establish the theoretical basis for BCA intervals. These intervals were grounded in an assumed relation between the skewness of the estimate and the relation between the variance of the estimate and the value of the parameter.

To establish this relation, consider a generic statistic,

$$\hat{\theta} = \theta + \dot{X}_{1,1} + \dot{X}_{2,1}\dot{X}_{2,2} + \cdots,$$

of the form of those described in Chapter 4. Without loss of generality, we assume $\mu_{X_{i,j}} = 0$ and $\mu_{X_{1,1}^2} = 1$. The series expansion of the variance of $\hat{\theta}$ is found by applying $\mathcal{T}_{\mu,\kappa[A]}$ to $\{\hat{\theta}, \hat{\theta}\}$ to yield a series with leading term equal to $1/n$.

The skewness of $\hat{\theta}$ is found by applying $\mathcal{T}_{\mu,\kappa[A]}$ to $\{\hat{\theta}, \hat{\theta}, \hat{\theta}\}$ to yield the series

$$\mathcal{T}_{\mu,\kappa[A]}[\{\hat{\theta}, \hat{\theta}, \hat{\theta}\}] \to n^{-2}\left(\mu_{X_{1,1}^3} + 6\mu_{X_{1,1}X_{2,1}}\mu_{X_{1,1}X_{2,2}}\right) + O(n^{-3}).$$

The series expansion of the estimate of b in (8.7) is found to be

$$n^{-1}\left((1/2)\mu_{X_{1,1}^3} + 2\mu_{X_{1,1}X_{2,1}}\mu_{X_{1,1}X_{2,2}}\right) + O(n^{-2}).$$

This covariance is almost proportional to the third cumulant of the estimate. The BCA correction is closely related to this skewness.

8.6 Some data analysis

Bootstrap methods were originally motivated by the need to compute cumulants for algebraically complicated statistics. The benefit is that, when the algebra is prohibitively complex, it can be replaced by simple simulation. However, the algebra can also be avoided by using the computer algebra tools discussed in this book. A comparison of methods is what is given here and three examples from Efron and Tibshirani (1993) are used for this purpose.

The simplest objective of this exercise is to produce results that are consistent with simulation and *vice versa*. An additional benefit includes the ability to easily place an analytic interpretation on the size of a simulation. Although problem dependent, simulations do not have to be excessively large to obtain results that agree with analytic bootstrap calculations to order n^{-1}. The ability to compute analytic cumulants and avoid sampling error, although appealing, is unlikely to result in the sudden abandonment of bootstrap simulation. The bootstrap can be applied to much more complicated situations than those considered here.

Correlation coefficient An example used effectively by Efron and Tibshirani to describe the basic bootstrap method involves LSAT and GPA scores from 82 law schools in the United States. Interest surrounds the correlation between the average LSAT score for entrants and the average GPA after one year. By considering a random sample of size (15), the authors show how the bootstrap distribution for the correlation coefficient mimics the true distribution based on the original 82 schools. Ideal bootstrap calculations in this case give values of -0.00466 and 0.132 for the bias and variance, respectively. In the case of the variance, the authors report a value of 0.132, indicating that second-order terms are captured by a simulation of size 3200.

Ratio of means The patch data arise from a test of bioequivalence for an old and a new patch, where both are designed to effect hormone levels in a person's blood stream. The parameter of interest is a ratio of means whose estimate is biased. The analytic bootstrap gives values of 0.0071 and 0.0101 for the bias and variance, respectively. The authors report values of 0.0043 and 0.105 for a simulation of size 400. Increasing the size of the simulation to 4000 results in values of 0.00679 and 0.0102. At least for the bias, it seems that a much larger simulation is needed to obtain results similar to the second-order analytic calculations.

Regression problem Here we consider a regression problem from Efron and Tibshirani concerning cell survival in the presence of varying doses of radiation. The data involve only one covariate and interest surrounds whether or not a quadratic term is appropriate. One issue concerning the authors is whether to bootstrap the covariates and response in pairs or to bootstrap only the residuals.

The disadvantage of the latter is that it requires the assumption of a possibly incorrect model while the former does not. However, bootstrapping in pairs treats the covariate as random, which is inappropriate in a controlled study such as this one.

The authors do well to point this out but neglect to provide bias estimates for regressions coefficients in this case. It is not surprising because for the conditional model being used that the bias estimates have no meaningful scientific interpretation. However, they are useful as indicators that we are not mimicking the real world appropriately and, as a consequence, bias estimates do not estimate 0 and variance estimates do not estimate the correct quantity either. None of this is as severe as resampling residuals under a large model departure.

For the linear and quadratic coefficients in the model, the bias–variance pairs are $(0.0177, 0.184)$ and $(-0.00222, 0.0225)$, respectively. The estimates of the standard deviation may be compared with 0.159 and 0.143 of Efron and Tibshirani, where the differences are likely due to sampling error. Perhaps this problem best exemplifies the usefulness of computer algebra as a means for studying the analytic properties of bootstrap methods.

8.7 Bibliographic notes

Efron and Tibshirani (1993) give a useful presentation of bootstrap methodology, concentrating on Monte Carlo applications. Hall (1992) provides additional theoritical information. The symbolic implementation described here was introduced in Andrews (1997). The BCA intervals of Efron and Tibshirani were based on the relation between the mean and variance of the estimate. Andrews (1999) discusses similar intervals based on the relation between the mean and the quantiles of the distribution of an estimate. Silverman and Young (1987) use computer algebra to evaluate criteria on which the decision to smooth a bootstrap distribution is based. Young and Daniels (1990) apply computer algebra to evaluate expressions in the assessment of bootstrap bias.

8.8 Problems

1. Compute the expansion of $\hat{\sigma}^2_{\hat{\theta}}$, the bootstrap estimate of the variance of $\hat{\theta} = \bar{X}$.

2. Compute the expansion of $\hat{\sigma}^2_t$, the bootstrap estimate of the variance of $t = (\hat{\theta} - \theta)/\sigma_{\hat{\theta}}$.

3. Compute the variances of the estimated variances in the previous two problems and show that the second is relatively more precise by an order of maginitude.

4. Consider the expansion the generic estimate $\hat{\theta} = \theta + \dot{X}_{1,1} + \dot{X}_{2,1}\dot{X}_{2,2} + \cdots$. Compute the bias of this estimate to order $O(n^{-1})$.

5. Bagged estimates, $\hat{\theta}_B$, proposed by Breiman and others are the average of estimates $\hat{\theta}$, calculated from bootstrap samples. Thus, they are Monte Carlo approximations of $E_{F_n}[\hat{\theta}]$. Because such estimates have been smoothed by

averaging, they often have smaller variance. However, the bias is greater. Show that, asymptotically, the bias of $\hat{\theta}_{\mathrm{B}}$ is twice that of $\hat{\theta}$.

6. The jackknife estimate of variance of an estimate $\hat{\theta}$ described in Mosteller and Tukey (1977) is based on pseudo-values p_i. These pseudo-values are computed from estimates $\hat{\theta}_{\hat{i}}$ based on all but the the ith observation,

$$p_i = n\hat{\theta} - (n-1)\hat{\theta}_{\hat{i}}.$$

The jackknife estimate of variance is s_p^2/n. Compare the asymptotic expansions of the bootstrap and jackknife estimates of variances of the generic statistic $\hat{\theta}$. To what order do they agree? Compare the variance of the difference of these estimates with that of the estimates themselves.

9
Sample surveys

9.1 Introduction

An introductory course in sample survey methodology is often riddled with tedious repetitive algebraic calculation of cumulants, and their estimates, for different designs. While the designs tend to be similar, the calculations are typically repeated in their entirety, obscuring their common structure. Symbolic computation is ideal in this context because the simple differences in design can be made explicit as inputs while the common features of each calculation are automated. Thus, the important distinctions among designs are emphasized and the redundant algebra eliminated.

The implementations of single-stage and multiple-stage designs are discussed in order, because additional complications arise for the latter. Algebra for single-stage designs is automated by simply extending the coefficient operator of Section 3.5.3 to encompass various sampling designs. In terms of transformations, this affects $\mathcal{T}_{\Delta[\mu],\Delta}$ and its inverse. Procedures for multiple-stage calculations exploit a repeated structure found in single-stage calculations.

Although the context of what follows is design-based, from a computational standpoint the inclusion of a model can be accommodated by simply adding another stage to the sampling design.

9.2 Extending the coefficient transformation

At each stage of a design, samples are drawn from a finite population either with or without replacement and with either equal or unequal probabilities. Sampling with replacement results in sums of IID random variables. The structure here is very similar to the bootstrap calculations of Chapter 8, especially if subsampling is permitted. Therefore, no broadening of operators is required.

Sampling without replacement presents an instance where the components of a sample sum are dependent. Here the traditional approach is to denote such a sum as

$$\sum_{i \in S} X_i,$$

where the index set S is a random variable and X_i is considered as fixed. While this may have its conceptual advantages, it is not convenient for computation.

Our operators act on products of sums whose arguments are random variables. To accommodate this, we replace the above sum with an equivalent form using indicator variables

$$\sum_{i=1}^{N} \{X_i I_{X_i}\},$$

where $I_{X_i} = 1$ if X_i is in the sample and 0 otherwise. Here X_i is still fixed but I_{X_i} is random.

Joint moments of the indicator variables are joint inclusion probabilities,

$$E[I_{X_i}] = P(\{i\} \subset S) = \pi_i,$$

$$E[I_{X_i} I_{Y_j}] = P(\{i, j\} \subset S) = \pi_{ij},$$

$$E[I_{X_i} I_{Y_j} I_{Z_k}] = P(\{i, j, k\} \subset S) = \pi_{ijk},$$

and these arise when computing the expected value of a disjoint sum, for example,

$$E\left[\sum_{i \neq j} X_i Y_j I_{X_i} I_{Y_j}\right] = \sum_{i \neq j} X_i Y_j \pi_{ij}.$$

Hence, we must extend the coefficient transformation, $\mathcal{C}$, of Section 3.5.3 for the computation of inclusion probabilities. In the special case of simple random sampling, characterized by every sample occurring with equal probability, the inclusion probability identified by the set J is $\pi_J = n^{(|J|)}/N^{(|J|)}$, where $|J|$ denotes the number of elements in the set and $n^{(r)}$ the factorial product $n(n - 1) \cdots (n - r + 1)$.

With this modification of $\mathcal{C}$, all the machinery for single-stage designs is in place.

9.3 Sampling designs

A list d is used to specify the sampling design of its argument as a list. The length of the list is the number of stages in the design. The coefficient operator uses the sampling design to compute the appropriate coefficient. The notation *srswr* and *ppswr* is used to denote simple random sampling and unequal probability sampling with replacement, respectively. Similarly, *srswor* and *ppswor* denote the above designs without replacement. In addition, we use the label *pop* to indicate when the entire (sub-) population is sampled and all inclusion probabilities are 1. For single-stage designs, such sums are treated as constants. However, for multiple-stage designs, such sums may need to be expanded in a calculation.

Stratified sampling In some cases an entire population may be partitioned into strata and samples drawn from each stratum. In the present context, this describes a two-stage design where at the first stage the entire population is sampled and at the second stage samples are drawn under some prescribed scheme. For example, the variable X can be specified to have this design by setting

$$d[X] = \{pop,\ srswor\},$$

where samples are drawn without replacement at the second stage.

Cluster sampling When units are sampled in groups or clusters, the design can be viewed as being in two stages where the entire sub-population is sampled at the second stage. So, here the design has the reverse structure of stratified sampling. For example, X can be specified to be the response for this design in the following way:

$$d[X] = \{srswor,\ pop\}.$$

9.4 Sampling notation for symbolic computation

Some further notation is required and this is summarized in Table 9.1. It is used for the remainder of this chapter.

In the case of multiple-stage designs, the notation for means is iterated; so, for example, a two-stage mean is written as $\widetilde{\widetilde{x}}$. In addition, the notation n and N is used for all stages of a design to represent the size of the sample and population at each stage. For example, a two-stage sample total may be written as $nn\widetilde{\widetilde{x}}$, where the inner n is indexed over the outer sum.

This iterated use of n and N is useful for efficient computation as it renders the behaviour of the coefficient transformation independent of the stage of the design. In addition, any number of stages can be considered without the need for further operators or notation. The other striking feature of the notation is the complete lack of subscripts for variables. This underlines the fact that while such appendages may be helpful for organizing calculations by hand, they are unnecessary for efficient computation.

The operators E, U, κ, and $(\cdot)_j$ from previous chapters are applicable in this context. Recall that the operator $\kappa_j[\cdot]$ computes the jth cumulant of its argument

Table 9.1. Sampling notation

Operator	Description	Example
$\overline{\Box}$	Population mean	$\overline{x}$
$\widetilde{\Box}$	Sample mean	$\widetilde{x}^2$
$\dot{\Box}$	Centred sample mean	$\dot{x} = \widetilde{x} - \overline{x}$
N	Population size	$\overline{x}$
n	Sample size	$\widetilde{x}$
$K_{\Box}$	Population k-statistic	$K_{x,y}$
$k_{\Box}$	Sample k-statistic	$k_{x,y}$

exactly, while $\kappa_{j,i}[\cdot]$ gives the expansion of the cumulant to order i. One additional operator, $\overline{x} : k_x$, writes means in terms of k-statistics.

9.5 Examples for single-stage designs

We demonstrate the procedures for single-stage designs by considering the efficient estimation of the population mean, $\overline{y}$, for a response y and throughout, x is assumed to be a covariate. Competing estimators include the sample mean, product estimator, ratio and regression estimators. Variance calculations reveal the hierarchy of these estimators. All calculations assume simple random sampling without replacement. The following examples show how cumulants may be computed for any design.

Cumulants of mean   The sample mean, $\widetilde{y}$, has cumulants that may be computed exactly:

$$d[y] = \{srswor\},$$

$$\kappa_1[\widetilde{y}] \longrightarrow \overline{y},$$

$$\kappa_2[\widetilde{y}] \longrightarrow \frac{(N-n)(\overline{y^2} - \overline{y}^2)}{n(N-1)},$$

$$(\overline{y} : k_y)[\kappa_2[\widetilde{y}]] \longrightarrow \frac{(N-n)K_{y,y}}{nN},$$

$$(\overline{y} : k_y)[\kappa_3[\widetilde{y}]] \longrightarrow \frac{(2n^2 - 3nN + N^2)K_{y,y,y}}{(nN)^2}.$$

Although $\widetilde{y}$ has the advantage of being unbiased, other estimators, that include information in the form of a covariate, x can have smaller mean square error. These estimators are unbiased to first order, in other words consistent, and hence it is enough to examine the leading term of the variance to see if we can improve upon $\widetilde{y}$.

Product estimator   The product estimator $p = \widetilde{x}\widetilde{y}/\overline{x}$ is a polynomial in sample means and hence its cumulants may be computed exactly. However, except for the mean, these expressions are large and not easily interpreted. Hence, we display the computed leading term in the expansion of the variance of the product estimator. Even though resorting to an asymptotic expression here is not necessary, it does give a more interpretable result:

$$p = \widetilde{x}\widetilde{y}/\overline{x},$$

$$(\overline{x} : k_x)[\kappa_1[p]] \longrightarrow K_y + \frac{(N-n)K_{x,y}}{NnK_x},$$

$$(\overline{x} : k_x)[\kappa_{2,0}[p]] \longrightarrow \frac{(N-n)(K_y^2 K_{x,x} + 2K_x K_y K_{x,y} + K_x^2 K_{y,y})}{NnK_x^2}.$$

As can be seen, the product estimator is unbiased to first order, but it only gives an estimator with much smaller variance if x and y are sufficiently negatively correlated.

Simple ratio estimator   The ratio estimator $r = \overline{x}\widetilde{y}/\widetilde{x}$ has cumulants that cannot be computed exactly and so we compute asymptotic approximations:

$$r = \overline{x}\widetilde{y}/\widetilde{x},$$

$$(\overline{x} : k_x)[\kappa_{1,0}[r]] \longrightarrow K_y,$$

$$(\overline{x} : k_x)[\kappa_{1,1}[r]] \longrightarrow 0,$$

$$(\overline{x} : k_x)[\kappa_{1,2}[r]] \longrightarrow \frac{(N-n)(K_y K_{x,x} - K_x K_{x,y})}{nN K_x^2},$$

$$(\overline{x} : k_x)[\kappa_{2,0}[r]] \longrightarrow \frac{(N-n)(K_y^2 K_{x,x} - 2K_y K_x K_{x,y} + K_x^2 K_{y,y})}{nN K_x^2}.$$

The ratio estimator is unbiased to first order and will improve upon the sample mean if x and y are sufficiently positively correlated.

Neither the product nor the ratio estimator have variances that are uniformly smaller than the variance of the sample mean. Their popularity has a historic basis: they only involve means and are easy to compute with mechanical calculators.

Regression estimator   Historically, the regression estimator was more complicated to compute because, for each sampling unit, the pair (x, y) must be matched in the calculation. This difficulty resulted in the continuing popularity of the product and ratio estimators long after it was proven that the regression estimator is uniformly better than the sample mean. This improvement is reflected in the expressions for the first two cumulants of the regression estimator r,

$$b = \frac{\widetilde{xy} - \overline{x}\widetilde{y}}{\widetilde{x^2} - \widetilde{x}^2},$$

$$r = \widetilde{y} + b(\overline{x} - \widetilde{x}),$$

$$(\overline{x} : k_x)[\kappa_{1,0}[r]] \longrightarrow K_y,$$

$$(\overline{x} : k_x)[\kappa_{1,1}[r]] \longrightarrow 0,$$

$$(\overline{x} : k_x)[\kappa_{1,2}[r]] \longrightarrow \frac{(N-n)(N-2)(K_{x,y} K_{x,x,x} - K_{x,x} K_{x,x,y})}{nN(N-1)K_{x,x}^2},$$

$$(\overline{x} : k_x)[\kappa_{2,0}[r]] \longrightarrow \frac{(N-n)(K_{y,y} K_{x,x} - K_{x,y}^2)}{nN K_{x,x}}.$$

If $K_{x,y}^2$, $K_{x,x}$ are both positive, the variance of r will always be less than $K_{y,y}$, at least to first order.

9.6 Multiple-stage designs

For multiple-stage designs, sums are indexed in more than one way. For example, in a design where enumeration areas and households are sampled, sums are indexed twice, once for each stage. Two complications arise for such sums. First, in an expected value calculation, products of sums must be expanded one stage at a time and, secondly, series expansions for means can have more than two terms. Using the methods of Chapters 3 and 4, we extend our procedures to calculations for multiple-stage designs.

9.6.1 OPERATORS FOR EXPECTATION

Consider the computation of the expected value of the product

$$\Pi = \sum_i \left\{ \sum_r X_{ir} \sum_s Y_{is} \right\} \sum_j \left\{ \sum_t Z_{jt} \right\},$$

which is of interest due to its repetitive structure: it is a product of sums where the terms in a sum may themselves be products of sums. The indices $i, j, \ldots$ are used for the first stage of a design and the indices $r, s, \ldots$ for the second stage. To compute the expected value of Π, it must be expanded one stage at a time. Letting $V_i = \sum_r Z_{ir}$ and $U_i = \sum_r X_{ir} \sum_r Y_{ir}$, we initially get

$$\Pi = \sum_i U_i V_i + \sum_{i \neq j} U_i V_j,$$

a sum over the full partition $\mathcal{P}_{\{U,V\}} = \{(UV), (U|V)\}$. Now the expected value operator can be expressed in nested form as $E = E_1 \circ E_2$, where $E_\bullet$ is an operator that is conditional on the previous stages of the design. For example, the behaviour of E_2 depends on how E_1 factored at the first stage. Unequal subscripts for the second term above indicate independent samples at the second stage conditional on the first stage. Hence, application of E results in

$$E[\Pi] = \sum_i \pi_i E_2[U_i V_i] + \sum_{i \neq j} \pi_{ij} E_2[U_i] E_2[V_j].$$

Determining the behaviour of E_2 now requires expanding the products $U_i V_i$, U_i, and V_j. These expansions are sums over the sets $\mathcal{P}_{\{X,Y,Z\}}$, $\mathcal{P}_{\{X,Y\}}$, and $\mathcal{P}_{\{X\}}$, respectively, where the full partitions computed at the second stage depend on the blocks of the partitions at the first stage. For example, the partition $(U|V)$ from the first stage has two blocks U and V, denoting the sets $\{X, Y\}$ and $\{Z\}$, respectively. At the second stage, the full partitions $\mathcal{P}_{\{X,Y\}}$ and $\mathcal{P}_{\{X\}}$ of these sets are computed in the calculation of $E_2[U]$ and $E_2[V]$.

In terms of the transformations of Chapter 3, the expansion in Section 9.6.1 is automated by $\mathcal{T}_{\Delta[\mu],A}$ and the action of E_2 by $\mathcal{T}_{A[\mu],A}$. The final result is then obtained by collapsing this expansion using $\mathcal{T}_{A[\mu],\Delta[\mu]}$. In this calculation the transformations $\mathcal{T}_{\Delta[\mu],A}$ and $\mathcal{T}_{A[\mu],\Delta[\mu]}$ are applied to the first stage and

$\mathcal{T}_{A[\mu],A}$ to the second. Hence, the ultimate transformation required is $\mathcal{T}^1_{A[\mu],\Delta[\mu]} \circ \mathcal{T}^2_{A[\mu],A} \circ \mathcal{T}^1_{\Delta[\mu],A}$, where superscripts indicate the stage of the design to which the transformation is applied.

The above structure is common to any product with any number of stages. When the sums have k stages, the transformation used is

$$\mathcal{T}^1_{A[\mu],\Delta[\mu]} \circ \cdots \circ \mathcal{T}^{k-1}_{A[\mu],\Delta[\mu]} \circ \mathcal{T}^k_{A[\mu],A} \circ \mathcal{T}^{k-1}_{\Delta[\mu],A} \circ \cdots \circ \mathcal{T}^1_{\Delta[\mu],A}.$$

Unbiased estimates are automated by the inverse transformation

$$\mathcal{T}^1_{A,\Delta} \circ \cdots \circ \mathcal{T}^{k-1}_{A,\Delta} \circ \mathcal{T}^k_{A,A[\mu]} \circ \mathcal{T}^{k-1}_{\Delta,A[\mu]} \circ \cdots \circ \mathcal{T}^1_{\Delta,A[\mu]}.$$

With this extension, all machinery for exact calculations for multiple-stage designs is in place.

9.6.2 EXPANSION OF A MEAN

For an estimator that is a smooth function of means, determining terms in its expansion ultimately requires computing terms in the expansion of a mean. In Section 4.2.6, the expansion of a mean was shown to have two terms. In our present context, means have multiple stages and each stage contributes two terms to the expansion. For example, a single-stage mean may be expanded as

$$\widetilde{x} = \overline{x} + \dot{x},$$

where $\dot{x} = \widetilde{x} - \overline{x}$. For a two-stage mean, we may use this expansion at the first stage and again at the second to get

$$\widetilde{\widetilde{x}} = \overline{\widetilde{x}} + \dot{\widetilde{x}} = \overline{\overline{x}} + \dot{\overline{x}} + \overline{\dot{x}} + \dot{x},$$

where $\overline{\dot{x}} = \overline{\widetilde{x}} - \overline{\overline{x}}$, $\dot{\overline{x}} = \overline{\widetilde{x}} - \overline{\overline{x}}$, and $\dot{x} = \widetilde{\widetilde{x}} - \overline{\widetilde{x}} - \overline{\widetilde{x}} + \overline{\overline{x}}$. Here the expansion has three terms

$$\widetilde{\widetilde{x}}_0 = \overline{\overline{x}}, \qquad \widetilde{\widetilde{x}}_1 = \dot{\overline{x}} + \overline{\dot{x}}, \qquad \widetilde{\widetilde{x}}_2 = \dot{x}.$$

The general rule used for the jth term in the expansion of $\widetilde{x}$ is

$$(\widetilde{x})_j = \sum_{i=0}^{j} (\widetilde{x_i})_{j-i},$$

where x itself may be a multiple-stage mean. The rule is the same as the rule for integer partitions, so a term in the expansion of multiple-stage mean can be identified by an integer partition. To avoid any redundant calculation, the partition should only involve 0s and 1s since the expansion at each stage involves only two terms. Adding this rule puts all the machinery for multiple-stage designs in place.

9.7 Examples for multiple-stage designs

The population mean can be estimated using data from a multiple-stage survey. Such surveys are conducted due to their cost-effectiveness when it is too expensive to sample randomly from the entire population. The tradeoff, a loss in efficiency for estimators, can be offset somewhat by designing the survey to take this into account.

Cumulants of a stratified mean In a stratified design, precision is gained through an appropriate choice of strata. Exact cumulants for a stratified mean may be computed as follows:

$$d[y] = \{pop, srswor\},$$

$$m = \frac{\overline{N\widetilde{y}}}{\overline{N}},$$

$$\kappa_1[m] \longrightarrow \frac{\overline{N\overline{y}}}{\overline{N}},$$

$$(\overline{x} : k_x)[\kappa_2[m]] \longrightarrow \left(\frac{\overline{N^2 K_{y,y}}}{n} - \overline{N K_{y,y}}\right)\bigg/ \overline{N}^2.$$

Cumulants of a clustered mean Clustering occurs when all units in a subpopulation are sampled, for example, all individuals in a dwelling. Cluster designs are often used when lists of sampling units are unavailable. So, a list of all dwellings may be available for a neighbourhood, but not a list of all residents. Efficiency is lost due to the probable positive correlation in the response for units in a cluster. Exact cumulants for a clustered mean may be computed as follows:

$$d[y] = \{srswor, pop\},$$

$$m = \frac{\widetilde{N\overline{y}}}{\overline{N}},$$

$$\kappa_1[m] \longrightarrow \frac{\overline{N\overline{y}}}{\overline{N}},$$

$$(\overline{x} : k_x)[\kappa_2[m]] \longrightarrow \frac{(N-n)K_{N\overline{y},N\overline{y}}}{nN\overline{N}^2}.$$

Cumulants of a two-stage mean Stratified and cluster sampling are not cases of multiple-stage designs since random sampling occurs only once. However, computation in these cases may be done using the methods for multiple-stage designs. In fact, it is necessary. This same technology can be used for general multiple-stage designs. Exact cumulants for a two-stage mean may be computed as follows:

$$d[y] = \{srswor, srswor\},$$

$$m = \frac{\widetilde{N\widetilde{y}}}{\overline{N}},$$

$$\kappa_1[m] \longrightarrow \frac{\overline{N\overline{y}}}{\overline{N}},$$

$$(\overline{x} : k_x)\,[\kappa_2[m]] \longrightarrow \frac{(N-n)K_{N\overline{y},N\overline{y}} + (N-1)\overline{(N/n-1)NK_{y,y}}}{n(N-1)\overline{N}^2}.$$

The interesting feature of the above expressions is the similarities with the expressions for the stratified and clustered designs. In the two-stage case all terms above contribute. This pattern also arises for ratio and regression estimators.

Ratio estimators   For each of the above designs, there is a competing ratio estimator that incorporates information through a covariate. In each case, the ratio estimator is unbiased to first order and will have smaller variance if the response is sufficiently positively correlated with the covariate. This may not be immediately apparent from the expression derived and some manipulation of the expression is required. General algorithms for this purpose are difficult because the form of the final expression is usually a matter of taste and will likely differ from user to user and problem to problem. In each of the following cases, the estimator is essentially the same—only the sampling design changes. These differences in design are made explicit through the use of the operators $\overline{\square}$ and $\widetilde{\square}$.

Clustered combined ratio estimator

$$d[y] = \{srswor, pop\}, \qquad d[y] = \{srswor, pop\},$$

$$r = \overline{N\overline{x}\,\widetilde{N\widetilde{y}}} / (\widetilde{N\widetilde{x}}\,\overline{N}),$$

$$\kappa_{2,0}\,[r] \longrightarrow \frac{(N-n)}{n(N-1)\overline{N}^2}$$

$$\times \left\{ \overline{N^2 K_x^2}\,\overline{NK_y}^2 / \overline{NK_x}^2 - 2\overline{NK_y}\,\overline{N^2 K_x K_y}/\overline{NK_x} + \overline{N^2 K_y^2} \right\}.$$

Stratified combined ratio estimator

$$d[y] = \{pop, srswor\}, \qquad d[x] = \{pop, srswor\},$$

$$r = \overline{N\overline{x}\,\widetilde{N\widetilde{y}}} / (\widetilde{N\widetilde{x}}\,\overline{N}),$$

$$(\overline{x} : k_x)\,\big[\kappa_{2,0}\,[r]\big] \longrightarrow \frac{\overline{NK_y}^2\,\overline{(N/n-1)NK_{x,x}}}{\overline{N}^2\,\overline{NK_x}^2} - 2\frac{\overline{NK_y}\,\overline{(N/n-1)NK_{x,y}}}{\overline{N}^2\,\overline{NK_x}}$$

$$+ \frac{\overline{(N/n-1)NK_{y,y}}}{\overline{N}^2}.$$

Two-stage combined ratio estimator

$$d[y] = \{srswor, srswor\}, \qquad d[x] = \{srswor, srswor\},$$

$$r = \overline{N\overline{x}}\widetilde{N\overline{y}}/(\widetilde{N\overline{x}N}),$$

$$(\overline{x} : k_x)\left[\kappa_{2,0}[r]\right]$$

$$\longrightarrow \frac{\overline{NK_y}^2\left\{4(N-n)\overline{N^2K_x^2} + (N-1)\overline{(N/n-1)NK_{x,x}}\right\}}{(N-1)n\overline{N}^2\overline{NK_x}^2}$$

$$- 2\frac{\overline{NK_y}\left\{4(N-n)\overline{N^2K_xK_y} + (N-1)\overline{(N/n-1)NK_{x,y}}\right\}}{(N-1)n\overline{N}^2\overline{NK_x}}$$

$$+ \frac{\left\{4(N-n)\overline{N^2K_y^2} + (N-1)\overline{(N/n-1)NK_{y,y}}\right\}}{(N-1)n\overline{N}^2}.$$

For the clustered combined ratio estimator, the entire population is sampled at the second stage and hence all interesting coefficients arise at the first stage of the design. This is not the case for the stratified combined ratio estimator as the entire population is selected at the first stage. Here all interesting coefficients arise at the second stage in the expression. The result for the two-stage combined ratio estimator involves both results above.

9.8 Bibliographical notes

The use of indicator variables to represent sample sums was first suggested by Cornfield (1944) and appears in several places in Cochran (1977). The methods of this chapter automate at least all the algebra in Cochran (1977) or any other sampling text at that level. Wishart (1952) showed that basic moment calculations under simple random sampling rely heavily on partitions. Some discussion of k-statistics and polykays for simple random sampling may be found in McCullagh (1987).

The computer algebra methods of this chapter are described in Stafford and Bellhouse (1997), and Bellhouse, Philips and Stafford (1997). Bellhouse (1999) uses the techniques of this chapter to demonstrate central limit theorems through an examination of cumulants for standardized statistics.

9.9 Problems

1. The terms in the expansion of a two stage mean, $\widetilde{\overline{x}}$, namely $\overline{\overline{x}}$, $\overline{\widetilde{x}}$, $\widetilde{\overline{x}}$, and $\dot{x}$ resemble those for main effects and interaction from a two-way analysis of variance. For example, $\dot{x} = \widetilde{\overline{x}} - \overline{\overline{x}} - \overline{\widetilde{x}} + \overline{\overline{x}}$. Why?

2. The analysis of contingency tables can involve statistics expressed in terms of sums with more than one index. Pearson's X^2 statistic is an example. Use

the methods of this chapter to construct a procedure for computing moments and cumulants of such statistics. You may start with Pearson's X^2 statistic, where

$$X^2 = N \left\{ \sum_{ij} \frac{a_{ij}^2}{s_i t_j} - 1 \right\}.$$

Here, a_{ij} denote cell counts, $N = \sum_{ij} a_{ij}^2$, and s_i, t_j the row and column totals, respectively. Extend the method to r-way contingency tables. Hint: the algorithm must expand sums, one index at a time; use the full partition operator $\mathcal{P}$. However, the sets that are partitioned for the second index are not the blocks of the partitions for the first index. That is, a full expansion is required at each stage. (See Stafford 1995.)

3. The analysis of variance can involve statistics expressed in terms of sums with more than one index. The standard F-tests are examples. Construct a procedure for computing moments and cumulants of such statistics. This procedure will be similar to the one from the previous question. Use the procedure to verify standard results under Gaussian assumptions and then repeat the calculations under some alternative distribution. The methods of Section 5.8 may be used for this purpose.

4. Use the methods of this chapter to derive the Horvitz–Thompson estimator and compute its first four cumulants.

5. By computing cumulants, use the methods of this chapter to demonstrate the central limit theorem for a standardized sum when the variance is assumed to be known or unknown.

6. By computing cumulants, use the methods of this chapter to demonstrate the central limit theorem for the ratio and regression estimators.

7. The modified ratio-product estimator is defined as

$$\omega \frac{\widetilde{y}\, \overline{x}\, \widetilde{z}}{\overline{z}\, \widetilde{x}}.$$

Find the value of ω that minimizes the mean square error of this estimator.

8. In some situations in a two-stage design, clusters may be sampled for some blocks and for others samples may be selected by some other means, for instance, simple random sampling. Modify the methods of this chapter to compute moments of a sample mean. Generalize this to cases where more than two sampling schemes are used.

9. Use the methods of this chapter to verify Tables 4.4 and 4.5 of McCullagh (1987).

10
Intersection matrices

10.1 Introduction

Cumulant calculations involve the application of an expected value operator E to products of constants, random variables or both. For the latter, grouping of random variables is required since E is a linear operator and the rule $E[a+bX] = a + bE[X]$ applies. In the scalar case, this is straightforward to implement since the commutative law of multiplication allows random variables to be grouped and separated from constants. However, the matter becomes complicated when we move to the multivariate case where the commutative law may not apply.

For example, consider the product $aXbY$, where a, b are constants, X, Y random variables, and all elements of the product are scalar. The expected value of this product is

$$E[aXbY] = abE[XY].$$

Here constants and random variables can be rearranged and separated. However, if all elements of the product are arrays, then we cannot assume that $a \cdot X \cdot b \cdot Y = a \cdot b \cdot X \cdot Y$. Hence, we cannot group the random variables and apply E. Such array products can be represented by graphs and this grouping would destroy the unique graphical structure of the product.

One of the two approaches adopted in this book is to abandon the '$\cdot$' notation in favour of another representation that permits the commutative law of multiplication but preserves the graphical structure of products. The summation convention is one such possibility, but as mentioned in Section 7.3, it requires brute-force procedures to simplify products in an expansion. Also, these procedures may fail to identify all equivalent products. Therefore, there is a need for an index-free representation that satisfies our requirements. Our solution involves intersection matrices which encapsulate the symmetries of a product and, as a result, its graphical structure. The matrices are unique up to permutations of rows and columns, which permits us to mimic the commutative law of multiplication.

10.2 Intersection matrices

10.2.1 GRAPHS AND SYMMETRY

The array products $a \cdot X \cdot b \cdot Y$ and $a \cdot b \cdot X \cdot Y$ can be written using the summation convention as $a_r X_{rs} b_{st} Y_t$ and $a_r b_{rs} X_{st} Y_t$ respectively, where a and Y are now

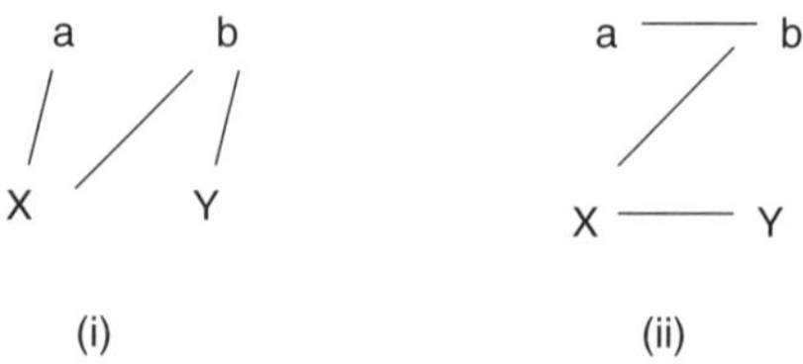

Fig. 10.1.   The products $a_r X_{rs} b_{st} Y_t$ and $a_r b_{rs} X_{st} Y_t$ have different graphical representations and are therefore distinct.

assumed to be vectors and X and b matrices. In the manner of Section 4.7, we can draw a graph (Figure 10.1) for each product, from which it is clear that the two products are distinct.

Now consider the products $a^r X_{sr} b^{st} Y_t$ and $a^r X_{rs} b^{st} Y_t$. They have the same graph but the two products are only equivalent if X is a symmetric matrix. That is, the graphs describe a unique product if the vertices of the graph are symmetric objects. For the graph symmetry means vertices are associated with sets of indices where, as in the case of the block of a partition, the order of the indices does not matter. If both X and b are symmetric, then for a product like $a^r b_{rs} X^{st} Y_t$, the number of indices shared by the elements of the product is the only important feature as it encapsulates the connectivity of the graph. The exact arrangement of subscripts and superscripts is irrelevant. Intersection matrices record the connectivity of a graph.

10.2.2 DEFINITION

In any computer algebra package, an intersection matrix is easy to define and manipulate. Consider a set S of k objects or indices $I_k = \{i_1, \ldots, i_k\}$. Let $p_1 = (b_1^1 | b_2^1 | \cdots | b_n^1)$ and $p_2 = (b_1^2 | b_2^2 | \cdots | b_m^2)$ be two partitions of S. The (i, j) entry of the intersection matrix $\mathcal{M}$ is

$$\mathcal{M}_{ij} = \#\{b_i^1 \cap b_j^2\}, \quad \forall i = 1, \ldots, n, \quad j = 1, \ldots, m.$$

Thus, $\mathcal{M}$ is an $n \times m$ matrix indexed by the blocks of each partition. For example, if $S = \{r, s, t, u\}$, $p_1 = (rs|t|u)$, $p_2 = (rt|su)$, then

$$\mathcal{M} = \begin{bmatrix} 1 & 1 \\ 1 & 0 \\ 0 & 1 \end{bmatrix}.$$

The intersection matrix encapsulates the connectivity of a graph since the (i, j)th entry is the number of lines in the graph that connect the vertices determining the blocks b_i^1 and b_j^2. For any array product written in index notation, the convention is that blocks of indices identify symmetries. The entry $\mathcal{M}_{ij}$ gives the order of the intersection of two blocks, while the position of the common indices does not matter. This is all the information required to identify the structure of the implicit sum and hence the array product.

For example, for the products $a^r X_{sr} b^{st} Y_t$ and $a^r X_{rs} b^{st} Y_t$, the set that is partitioned is $S = \{r, s, t\}$, the set of indices that are duplicated in the representation of the product. The partitions of interest are $p_1 = (rs|t)$ and $p_2 = (r|st)$, one given by the subscripts and the other by the superscripts. Blocks are identified by the different terms in each product. So here the condition that order within a block is irrelevant is consistent with the symmetry of X and b. Under such symmetry, the two products are equivalent and hence have the same intersection matrix, namely

$$\mathcal{M} = \begin{bmatrix} 1 & 1 \\ 0 & 1 \end{bmatrix},$$

which identifies the connectivity of the graph on the left of Figure 10.1.

For another example, suppose

$$T_1 = U^{rs} L_{st} U^{tu} L_{uv} U^{vw} L_{rw} \quad \text{and} \quad T_2 = U^{rs} L_{su} U^{ut} L_{tv} U^{vw} L_{rw},$$

where U and L are both symmetric. While it is easy to see that in the second product we have simply transposed t with u and hence the products are equivalent, a computer will treat these products as distinct. For T_1 and T_2, the set that is partitioned is $S = \{r, s, t, u, v, w\}$ and for T_1 the partitions of interest are

$$L_{st} L_{uv} L_{rw} \implies p_1 = (rw|st|uv),$$
$$U^{rs} U^{tv} U^{uw} \implies p_2 = (rs|tv|uw).$$

The intersection matrix in this case is

$$\mathcal{M} = \begin{bmatrix} 1 & 0 & 1 \\ 1 & 1 & 0 \\ 0 & 1 & 1 \end{bmatrix}.$$

Computing the intersection matrix for T_2 gives the same result and hence the products are equivalent.

For algorithms that return objects written in the index notation, the intersection matrix provides an effective means of eliminating redundant terms and simplifying expressions.

10.2.3 THE COMMUTATIVE LAW OF MULTIPLICATION

The use of index notation allows array products to be represented componentwise as a single term in an implicit sum. Thus, array products are represented as scalar products and the commutative law of multiplication applies. For intersection matrices, we have arbitrarily let subscripts determine the rows and superscripts the columns. The commutative law means the order of the blocks of each partition is irrelevant and they can be re-arranged without affecting the value of the term. In other words, intersection matrices are unique up to permutations of rows and columns.

Fig. 10.2. The two graphs represent the same product. A canonical ordering of the intersection matrix is required to avoid a graph isomorphism problem.

For example, T_2 may be written as $U^{rs}L_{su}U^{ut}L_{tv}U^{vw}L_{rw}$ or $U^{ut}L_{su}U^{rs}L_{tv}U^{vw}L_{rw}$, giving the two matrices

$$\begin{bmatrix} 1 & 1 & 0 \\ 0 & 1 & 1 \\ 1 & 0 & 1 \end{bmatrix} \quad \text{and} \quad \begin{bmatrix} 1 & 1 & 0 \\ 1 & 0 & 1 \\ 0 & 1 & 1 \end{bmatrix},$$

corresponding to the graphs in Figure 10.2.

Now two distinct graphs identify the same product and there appears to be a graph isomorphism problem. However, this problem can be avoided by pursuing a canonical ordering of the matrix. One strategy involves sorting the rows and columns, so they observe the separation of constants and random variables. That is, we mimic the commutative law of multiplication in our favour.

10.3 Four examples

The following are examples of calculations that can be automated through the use of intersection matrices. In each case, the use of the intersection matrix is highlighted, while the full implementation is left as an exercise. The third example presents a slight variation of the definition of $\mathcal{M}$ and, as a result, shows the breadth of application of such operators.

10.3.1 COMPLEMENTARY SET PARTITIONS AND EQUIVALENCE CLASSES

Consider the covariance of two quadratic forms $U = a^{ij}X_iX_j$ and $V = b^{ij}X_iX_j$, where only ordinary cumulants are available. We have

$$Cov(U, V) = a^{ij}b^{kl}Cov(X_iX_j, X_kX_l)$$

$$= a^{ij}b^{kl}\kappa_{ij,kl}$$

$$= a^{ij}b^{kl}(\cdots + \kappa_j\kappa_l\kappa_{ik} + \cdots + \kappa_j\kappa_k\kappa_{il} + \cdots),$$

where $\kappa_{ij,kl}$ is a generalized cumulant identified by the partition $(ij|kl)$. The last expression gives $\kappa_{ij,kl}$ in terms of ordinary cumulants, an operation automated in Section 3.8.1. This last expression is also a sum over the complementary set partition (McCullagh 1987) of $(ij|kl)$.

If a_{ij} and b_{kl} are symmetric matrices, the terms $a^{ij}b^{kl}\kappa_j\kappa_l\kappa_{ik}$ and $a^{ij}b^{kl}\kappa_j\kappa_k\kappa_{il}$ are equivalent since we are free to exchange the indices k and l in either term. In both cases, the intersection matrix is the same, namely

$$\begin{bmatrix} 1 & 0 \\ 0 & 1 \\ 1 & 1 \end{bmatrix},$$

which identifies the equivalence. In fact, any term in the expression for $\kappa_{ij,kl}$ with the above intersection matrix makes the same contribution to the sum and hence is said to belong to the same equivalence class. The simplification problem of interest here then consists of reducing a complementary set partition to a set of equivalence classes, and intersection matrices can be used for this purpose.

The operator $\mathcal{E}q\mathcal{C}l$ gives the complementary set partition for a partition p as a sum over a set of equivalence classes. Each class is identified by a matrix given as a nested list. For example,

$$\mathcal{E}q\mathcal{C}l[(i\,|\,jk)] \longrightarrow \{\{1,2\}\} + 2\{\{0,1\},\{1,1\}\},$$

$$\mathcal{E}q\mathcal{C}l[(ij\,|\,kl)] \longrightarrow \{\{2,2\}\} + 2\{\{0,1\},\{2,1\}\} + 2\{\{1,0\},\{1,2\}\}$$

$$+ 2\{\{1,1\},\{1,1\}\} + 4\{\{0,1\},\{1,0\},\{1,1\}\}.$$

In the first example, both partitions $(s\,|\,rt)$ and $(t\,|\,rs)$ belong to the second equivalence class because the intersection matrix for the pairs $(r\,|\,st)$, $(s\,|\,rt)$, and $(r\,|\,st)$, $(t\,|\,rs)$ is the same, namely $\{\{0,1\},\{1,1\}\}$. The operator $\mathcal{E}q\mathcal{C}l$ may be used to automate cumulant calculations such as the one above or any other calculation involving complementary set partitions, such as the one in the next example.

10.3.2 CUMULANTS OF k-STATISTICS

Here we compute cumulants of k-statistics using intersection matrices, an approach that differs from that of Section 3.11.3. Consider computing the covariance of two k-statistics: a sample mean $\bar{x}$ and a sample covariance s_{yz}. We may write $\bar{x} = \phi^i x_i$ and $s_{yz} = \phi^{ij} y_i z_j$, where $\phi^i = 1/n$, and

$$\phi^{ij} = \begin{cases} 1/n, & i = j, \\ -1/[n(n-1)], & i \neq j. \end{cases}$$

By vectorizing both sums, we see that the covariance is

$$Cov[\bar{x}, s_{yz}] = \phi^i \phi^{jk} Cov[x_i, y_j z_k]$$

$$= \phi^i \phi^{jk} \kappa_{i,jk}$$

$$= \phi^i \phi^{jk} \{\kappa_{i,j,k} + \kappa_j \kappa_{i,k} + \kappa_k \kappa_{i,j}\},$$

where the final expression is a sum over the complementary set partition of $(i\,|\,jk)$. Now, for a sequence of independent and identically distributed random variables,

Fig. 10.3.   The unique graphs for the three terms in the expansion of $Cov[\bar{x}, s_{yz}]$. The graph on the right identifies two terms.

$\kappa_{i,j,k} = 0$ unless $i = j = k$, and $\kappa_{i,j} = 0$ unless $i = j$. Thus, the expression may be written in terms of Kronecker deltas:

$$Cov[\bar{x}, s_{yz}] = \phi^i \phi^{jk} \{\delta_{ijk}\kappa_{x,y,z} + \delta_j \delta_{ik}\kappa_y \kappa_{x,z} + \delta_k \delta_{ij}\kappa_z \kappa_{x,y}\}.$$

Part of the final evaluation of this expression involves simplifying the superscripts using the Kronecker deltas and simultaneously identifying equivalent terms. An intersection matrix can be used for this purpose. For example, both $\phi^i \phi^{jk} \delta_j \delta_{ik}$ and $\phi^i \phi^{jk} \delta_k \delta_{ij}$ become $\sum_{i,j} \phi^i \phi^{ij}$ (Figure 10.3(b)), as is evident from the intersection matrix in each case:

$$\begin{bmatrix} 0 & 1 \\ 1 & 1 \end{bmatrix}.$$

Other steps in this calculation are automated using the methods of Chapter 3, where it is important to realize that expressions like $\sum_{i,j} \phi^i \phi^{ij}$ never arise.

The following are examples of cumulant calculations for k-statistics using an operator denoted as $\mathcal{K}$:

$$\mathcal{K}[k_x, k_{y,z}] \longrightarrow \frac{\kappa_{x,y,z}}{n},$$

$$\mathcal{K}[k_{x,y}, k_{z,w}] \longrightarrow \frac{\kappa_{x,y,z,w}}{n} + \frac{2\kappa_{x,z}\kappa_{y,w}}{n-1}.$$

Rather than resorting to the present approach, cumulants of k-statistics can be computed directly using the methods of Chapter 3. Each implementation has advantages over the other, making neither uniformly more efficient.

10.3.3   MULTIPLE-PHASE DESIGNS

The next example shows how slight variations in the definition of a matrix can provide more general algorithms. Other examples of this idea are given as exercises. The example extends the methods of Chapter 9 to multiple-phase designs.

In sampling any stage of a design can involve phases where, at each phase, units are sub-sampled. Automating cumulant calculations for such designs requires identifying how an expected value operator E should behave. An indicator matrix can be used for this purpose and, for the sake of exposition, we consider only a single-stage design.

Consider computing the expected value of the product

$$\Pi = \sum X \sum Y \sum Z,$$

where the Y's result from a first phase, the X's from a second phase and the Z's from a third. The notation π^r is used to denote an inclusion probability at the rth phase. Following the strategy of Chapter 3, Π can be expanded where one term in the result is $t = \sum_{i \neq j \neq k} X_i Y_j Z_k$, which is identified by the partition $(X|Y|Z)$. Regarding E as being the composition of several conditional operators (one for each stage), the expected value of t is

$$\sum_{i \neq j \neq k} \left\{ \pi^1_{ijk} \pi^2_{ik} \pi^3_k X_i Y_j Z_k \right\},$$

where π^2_{ik} arises because both X and Z involve at least two phases. For example, if simple random sampling is used at each phase the result becomes

$$\sum_{i \neq j \neq k} \left\{ \frac{n^1(n^1 - 1)(n^1 - 2)}{N(N - 1)(N - 2)} \times \frac{n^2(n^2 - 1)}{n^1(n^1 - 1)} \times \frac{n^3}{n^2} \times X_i Y_j Z_k \right\}$$

or rather

$$\sum_{i \neq j \neq k} \left\{ \frac{n^3(n^2 - 1)(n^1 - 2)}{N(N - 1)(N - 2)} \times X_i Y_j Z_k \right\},$$

where n^r denotes the size of the sample at the rth phase.

From a computer algebra perspective, the implementation issue is to recognize from the partition $(X|Y|Z)$ that the expected value of t involves the inclusion probability $\pi^1_{ijk} \pi^2_{ik} \pi^3_k$. Here we define a matrix where each entry indicates whether a block of a partition contains a unit that arises from a particular phase, that is

$$\mathcal{M}_{ij} = \begin{cases} 1, & \text{if the } i\text{th block has a unit with } j \text{ phases,} \\ 0, & \text{otherwise.} \end{cases}$$

For our example,

$$\mathcal{M} = \begin{bmatrix} 1 & 1 & 0 \\ 1 & 0 & 0 \\ 1 & 1 & 1 \end{bmatrix},$$

where the pattern of 0's and 1's in the rth column indicates how π^r should be subscripted. Another example helps clarify the method. For the term identified by the partition $(XY|Z)$, the indicator matrix is

$$\mathcal{M} = \begin{bmatrix} 1 & 1 & 0 \\ 1 & 1 & 1 \end{bmatrix},$$

yielding $\sum_{i \neq j} \pi^1_{ij} \pi^2_{ij} \pi^3_j X_i Y_i Z_j$.

10.3.4 MORE LIKELIHOOD CALCULATIONS

Continuing the likelihood example of Chapter 4, and following the notation given there, we may compute moments by abandoning the '$\cdot$' notation in favour of the summation convention. Expressions that result may be simplified through the use of Bartlett's identities (Section 7.3.4). However, this can result in redundant terms in an expression. In Chapter 7, we showed how these could be eliminated by inelegant brute-force procedures. Here we use intersection matrices.

For example, two terms in the expected value of $z_e z_f z_g z_h$ are $I_{e,f} I_{g,h}$ and $I_{e,g} I_{f,h}$. Thus, upon simplification, two terms in the expected value of $K^{ae} K^{bf} K^{cg} K^{dh} I_{abcd} z_e z_f z_g z_h$ are $K^{ab} K^{cd} I_{abcd}$, and $K^{ac} K^{bd} I_{abcd}$ which are equivalent except that the computer does not recognize this. Here the intersection matrices are the same and the equivalence recognized.

Using a combination of operators, expressions for the bias of $\hat{\theta}$ and the expected value of $\overline{w(\theta)}$ can be derived where no redundant terms appear in the result:

$$\mathcal{AE}_2[\kappa_{\hat{\theta}}] \longrightarrow I_{u,st} K^{ut} K^{rs} + \frac{I_{stu} K^{rs} K^{tu}}{2},$$

$$\mathcal{AE}_2[\kappa_{w(\theta)}] \longrightarrow p + \frac{1}{n}\Big\{ I_{r,s,tu} K^{ru} K^{st} - I_{uvw} I_{r,s,t} K^{rw} K^{sv} K^{tu}$$

$$+ I_{r,stu} K^{ru} K^{st} + I_{rs,tu} K^{rt} K^{su} + \frac{I_{rstu} K^{rs} K^{tu}}{4}$$

$$- I_{r,st} I_{u,vw} K^{rw} K^{sv} K^{tu} - 2 I_{uvw} I_{r,st} K^{rw} K^{su} K^{tv}$$

$$- \frac{I_{rst} I_{uvw} K^{ru} K^{sv} K^{tw}}{2} - \frac{I_{rst} I_{uvw} K^{rw} K^{st} K^{uv}}{4}$$

$$- I_{uvw} I_{r,st} K^{rt} K^{sw} K^{uv} - I_{r,st} I_{u,vw} K^{rt} K^{sw} K^{uv} \Big\},$$

where $\mathcal{AE}_i$ denotes the asymptotic expansion to order $O(n^{-i/2})$.

10.4 An alternative implementation

The above approach is inefficient as redundant terms are generated only to be eliminated later by an operator that uses an intersection matrix. An alternative approach is proposed. When products have graph representations, their manipulation using symbolic computation is analogous to mapping one graph to another to another, and so forth. In simple calculations, where, for instance, symmetry abounds, this mapping can be mimicked by matrix operations.

For example, consider calculating the expected value of $K^{rs} I_{stu} K^{tv} K^{uw} z_v z_w$:

$$E[K^{rs} I_{stu} K^{tv} K^{uw} z_v z_w] = K^{rs} I_{stu} K^{tv} K^{uw} E[z_v z_w]$$

$$= -K^{rs} I_{stu} K^{tv} K^{uw} I_{vw} + K^{rs} I_{stu} K^{tv} K^{uw} I_v I_w$$

$$= K^{rs} I_{stu} K^{tu}.$$

The second line results from the full partition of the set $\{z_v, z_w\}$, where the second term is actually 0 given $I_v = I_w = 0$. The third line results from the application of Bartlett's second identity to the first term on the second line.

Each of these steps can be viewed as manipulation of the original intersection matrix for $\mathcal{T}$. Here we reproduce the the above calculation by giving the appropriate intersection matrix for each term

$$\begin{bmatrix} 1 & 1 & 1 \\ 0 & 1 & 0 \\ 0 & 0 & 1 \end{bmatrix} \rightarrow \begin{bmatrix} 1 & 1 & 1 \\ 0 & 1 & 0 \\ 0 & 0 & 1 \end{bmatrix}$$

$$\rightarrow \begin{bmatrix} 1 & 1 & 1 \\ 0 & 1 & 1 \end{bmatrix} \quad \begin{bmatrix} 1 & 1 & 1 \\ 0 & 1 & 0 \\ 0 & 0 & 1 \end{bmatrix}$$

$$\rightarrow \begin{bmatrix} 1 & 2 \end{bmatrix}.$$

From this we can see that the process of computing the expectation of $K^{rs} I_{stu} K^{tv} K^{uw} z_v z_w$ and simplifying the result is a matrix operation where:

For the matrices in the first line, the second and third rows indicate structure for the two random variables $z_v z_w$. Application of the expected value operator amounts to computing the full partition of these two rows, where, in the result, rows that belong to the same block are added together giving the second line.

The number of non-zero entries in a row indicates the order of the derivative of the log-likelihood. A row with one non-zero entry represents the expected value of the score function and is hence a term with value zero. Thus, the second matrix in the second line can be dropped.

However, the first matrix has a second row that identifies the expected Fisher's information $-I_{vw}$. The cancellation $-I_{vw} K^{vw} = 1$ is accomplished by the elimination of this row and the addition of the columns which had a 1 in their entry for the eliminated row. This results in the matrix of the third line.

The final matrix identifies the correct result $K^{rs} I_{stu} K^{tu}$. Figure 10.4 depicts the calculation giving the appropriate graph for each term. A general implementation of this algorithm is left as an exercise (for the authors too).

10.5   Bibliographical notes

The methods of this chapter appear in Stafford (2000). Intersection matrices are discussed in McCullagh (1987) in connection with complementary set partitions and their reduction to a set of equivalent classes. Automating the recognition of equivalent terms written in the summation convention is a problem that is not unique to statistics. It also arises in mathematics and physics, particularly general relativity where tensor arithmetic is prevalent. Considerable effort has been made to construct a system for automatic tensor arithmetic. An example of such a system is given in Lee (1997). Within this system, simplification procedures resort to

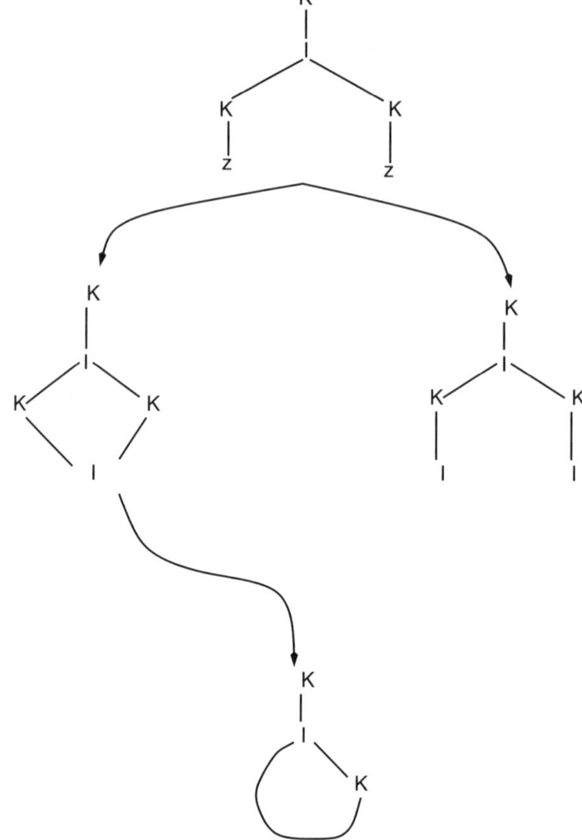

Fig. 10.4. A graphical summary of the computation of the expectation of $K^{rs}I_{stu}K^{tv}$ $K^{uw}z_v z_w$

canonical representation of indices and swapping of indices in a manner similar to Section 7.3.1. These procedures take advantage of symmetries in tensors by another technique. The elimination of all redundant terms is not guaranteed, though the simplification problem here is much more difficult than the ones we have considered. A useful discussion of this issue may be found in McKay (1991).

Here the graphical representation used has been unique due to the simplicity of the problems considered. Symmetry of arrays, or a problem-dependent canonical ordering, allows full simplification. There has been no need to check the equivalence of graphs by, say, 'swapping' vertices.

One could consider more difficult simplification problems such as the utilization of Bartlett identities (Section 7.3.4), which equate to zero a sum over the full partition of a set of indices. Analogous Bianchi identities arise in general relativity (Lee, 1997). In either case, the problem of how to best implement such identities to obtain the simplest expression possible has, to date, been approached in brute-force manner by exhaustively considering all possible substitutions. However, the expression to be simplified may have redundant terms present and so it is difficult to imagine that the optimal substitution will be found. The appeal to the graphical connection may offer a more elegant implementation. Just this problem is considered in Kendall (1994) using a different approach.

10.6 Problems

1. The following definition generalizes the notion of an intersection matrix: for two sets $S_1 = (s_1^1|s_2^1|\cdots|s_n^1)$ and $S_2 = (s_1^2|s_2^2|\cdots|s_m^2)$ whose blocks are subsets of a common set S, $\mathcal{M}$ has entry

$$\mathcal{M}_{ij} = g\{f(s_i^1), h(s_j^2)\}, \quad \forall i = 1, \ldots, n, \ \ j = 1, \ldots, m,$$

   where g, f, and h are arbitrary functions. The matrices defined in this chapter are special cases of the above definition. This definition can be extended to more than two sets, in which case our intersection matrix becomes a higher-order array.

   Within the context of factorial models, it may be useful if the computer automatically recognized the hierarchy. Let $\oplus$ and $\otimes$ denote additive effects and effects with interaction, respectively. For example, $A \oplus B$ denotes the model with main effects A and B but no interaction; $A \otimes B$ denotes the same model where an interaction term is included. A matrix can be defined to identify $A \oplus B$ as a sub-model of $A \otimes B$ if it performs a pairwise check of each additive component of each model.
   Define a matrix that performs such checks for two models where, in general, for two models

$$m_1 = s_1^1 \oplus s_2^1 \oplus \cdots \oplus s_n^1 \quad \text{and} \quad m_2 = s_1^2 \oplus s_2^2 \oplus \cdots \oplus s_m^2.$$

   An example of this notation is $m_1 = A \oplus B$ and $m_2 = A \otimes B$, where $s_1^1 = A$, $s_2^1 = B$, and $s_1^2 = A \otimes B$.

2. For factorial models of the type considered in the previous question, the full distributive lattice for a set of factors $\{A_1, \ldots, A_k\}$ is the collection of all possible models using these factors. Construct an algorithm that generates the full distributive lattice for such a set.

3. Use the methods of this chapter to verify Table 1 and Table 2 in the appendix of McCullagh (1987).

4. Implement the algorithm of Section 10.4.

References

Andrews, D.F. (1997). The analytic jackknife. *The Practice of Data Analysis*. Wiley: New York.

Andrews, D.F. (1999). A note on the computation of expansions of the signed root log-likelihood ratios. *Technical Report*, Department of Statistics, University of Toronto.

Andrews, D.F. (1999). Confidence limits based on quantile stabilizing transformations. *Technical Report*, Department of Statistics, University of Toronto.

Andrews, D.F. (2000). Asymptotic expansions of moments and cumulants. *Statistics and Computing*, in press.

Andrews, D.F. and Feuerverger, A. (1995). Generalized saddlepoint approximations. *Technical Report*, Department of Statistics, University of Toronto.

Andrews, D.F. and Stafford, J.E. (1993). Tools for the symbolic computation of asymptotic expansions. *Journal of the Royal Statistical Society*, **B55**, 613–628.

Andrews, D.F. and Stafford, J.E. (1998). Iterated full partitions. *Statistics and Computing*, **8**, 189–192.

Aspin, A.A. (1948). An examination and further development of a formula arising in the problem of comparing two means. *Biometrika*, **35**, 88–96.

Barndorff-Nielsen, O.E. (1983). On a formula for the distribution of the maximum likelihood estimator. *Biometrika*, **70**, 343–365.

Barndorff-Nielsen, O.E. (1986). Inference on full or partial parameters based on the standardized signed log-likelihood ratio. *Biometrika*, **73**, 307–322.

Barndorff-Nielsen, O.E. and Blaesild, P. (1986). A note on the calculation of Bartlett adjustments. *Journal of the Royal Statistical Society*, **B48**, 351–358.

Barndorff-Nielsen, O.E. and Cox, D.R. (1979). Edgeworth and saddlepoint approximations with statistical applications (with discussion). *Journal of the Royal Statistical Society*, **B41**, 279–312.

Barndorff-Nielsen, O.E. and Cox, D.R. (1989). *Asymptotic Techniques for Use in Statistics*. Chapman and Hall: London.

Barndorff-Nielsen, O.E. and Cox, D.R. (1994). *Inference and Asymptotics*. Chapman and Hall: London.

Bellhouse, D.R. (1999). The central limit theorem under simple random sampling. *Technical Report*, Department of Statistics, University of Western Ontario.

Bellhouse, D.R., Philips, R., and Stafford, J.E. (1997). Symbolic operators for multiple sums. *Journal of Computational Statistics and Data Analysis*, **24**, 443–454.

Brillinger, D.R., (1975) *Time Series: Data Analysis and Theory*. Holt Rinehart and Winston: New York.

Cochran, W.G. (1977). *Sampling Techniques*. Wiley: New York.

Cornfield, J. (1944). On samples from finite populations. *Journal of the American Statistical Association*, **39**, 236–239.

Cox, D.R. and Reid, N. (1987). Parameter orthogonality and approximate conditional inference. *Journal of the Royal Statistical Society*, **B49**, 1–39.

Daniels, H.E. (1954). Saddlepoint approximations in statistics. *Annals of Mathematical Statistics*, **25**, 631–650.

Daniels, H.E. (1987). Uniform approximations for tail probabilities. *International Statistical Review*, **55**, 37–48.

Daniels, H.E. and Young, G.A. (1991). Saddlepoint approximations for the studentized mean, with an application to the bootstrap. *Biometrika*, **78**, 169–179.

Davenport J.H., Siret Y., and Tournier, G. (1988). *Computer Algebra, Systems and Algorithms for Algebraic Computation*. Academic Press.

Davison, A.C. and Stafford, J.E. (1998). A comparison of adjusted score functions. *Canadian Journal of Statistics*, **26**, 139–148.

DiCiccio, T.J. (1984). On parameter transformations and interval estimation. *Biometrica*, **71**, 477–485.

DiCiccio, T.J., Hall, P., and Romano, J. (1991). Empirical likelihood is Bartlett-correctable. *Annals of Statistics*, **19**, 1053–1061.

Diciccio, T.J. and Martin, M.A. (1991). Approximation of marginal tail probabilities for a class of smooth functions with applications to Bayesian and conditional inference. *Biometrika*, **78**, 891–902.

DiCiccio, T.J. and Stern, S.E. (1994). Frequentist and Bayesian Bartlett correction of test statistics based on adjusted profile likelihood. *Journal of the Royal Statistical Society*, **B56**, 397–408.

Easton, G.S. and Ronchetti, E. (1986). General saddlepoint approximations with applications to L statistics. *Journal of the American Statistical Association*, **81**, 420–430.

Efron, B. and Tibshirani, R. (1993). *An Introduction to the Bootstrap*. Chapman and Hall: New York.

Field, C. and Ronchetti, E. (1990). *Small Sample Asymptotics*. Institute of Mathematical Statistics Lecture Notes, **13**: Hayward, California.

Firth, D. (1993). Bias reductions of maximum likelihood estimates. *Biometrika*, **80**, 27–38.

Fisher, R.A. (1921). On the probable error of a coefficient of correlation deduced from a small sample. *Metron*, **1**, 3–32.

Fisher, R.A. (1928). Moments and product moments of sampling distributions. *Proceedings London Mathematics Society*, **30**, 199–238.

Hall, P. (1992) *The Bootstrap and Edgeworth Expansion*. Springer-Verlag: New York.

Heller, B. (1991) *MACSYMA for Statisticians*. Wiley: New York.

James, G.S. (1958). On moments and cumulants of systems of statistics. *Sankhya*, **20**, 1–30.

Jing, B.Y., Feuerverger, A., and Robinson, J. (1994). On the bootstrap saddlepoint approximations. *Biometrika*, **81**, 211–215.

Kendall, M.G. and Stuart, A. (1977). *The Advanced Theory of Statistics*. 4th edition. Griffin: London.

Kendall, W.S. (1993). Computer algebra in probability and statistics. *Statistica Neerlandica*, **47**, 9–25.

Kendall, W.S. (1994). Computer Algebra and Yoke Geometry I: when is an expression a tensor? *Technical Report*, Department of Statistics, Warwick University.

Lawley, D.N. (1956). A general method for approximating to the distribution of the likelihood ratio criteria. *Biometrika*, **43**, 295–303.

Lee, J. (1997). Ricci: A Mathematica package for doing tensor calculations in differential geometry. *Technical Report*, Department of Mathematics, University of Washington.

Leonov, V.P. and Shiryaev, A.M. (1959). On a method of calculation of semi-invariants. *Theory of Probability and Applications*, **4**, 319–329.

Lugannani, R. and Rice, S. (1980). Saddlepoint approximation for the distribution of a sum of independent random variables. *Advanced Applications in Probability*, **12**, 475–490.

McCullagh, P. (1987). *Tensor Methods in Statistics*. Chapman and Hall: London.

McCullagh, P. and Tibshirani, R. (1990). A simple method for the adjustment of profile likelihoods. *Journal of the Royal Statistical Society*, **B52**, 325–344.

McCullagh, P. and Wilks, A. (1988). Complementary set partitions. *Proceedings Royal Society London*, **A415**, 347–362.

McKay, B. (1991). Nauty user's guide. *Technical Report*, Department of Computer Science, Australian National University.

Mosteller, F. and Tukey, J.W. (1977). *Data Analysis and Regression*. Addison Wesley: Reading, Massachusetts.

Peers, H.W. and Iqbal, M. (1985). Asymptotic expansions for confidence limits in the presence of nuisance parameters, with applications. *Journal of the Royal Statistical Society*, **B47**, 547–554.

Pistone, G., Riccomagno, E., and Wynn, H.P. (1999). Polynomial encoding of discrete probability using Gröbner bases. *Proceedings of the International Statistical Institute*.

Pistone, G., Riccomagno, E., and Wynn, H.P. (2000). Gröbner bases and factorisation in discrete probability and Bayes. *Statistics and Computing*, in press.

Reid, N. (1988). Saddlepoint methods in statistical inference (with discussion). *Statistical Science*, **3**, 213–227.

Silverman, B.W. and Young, A.E. (1987). The bootstrap: to smooth or not to smooth? *Biometrika*, **74**, 469–479.

Smith, B. and Field, C. (2000). Symbolic cumulant calculations for frequency domain time series. *Statistics and Computing*, in press.

Snedecor, G.W. and Cochrane, W.G. (1967). *Statistical Methods*. Iowa State University Press: Ames Iowa.

Stafford, J.E. (1994). Automating the partition of indexes. *Journal of Computational and Graphical Statistics*, **3**, 249–260.

Stafford, J.E. (1995). Exact cumulant calculations for Pearson X^2 and Zelterman statistics for r-way contingency tables. *Journal of Computational and Graphical Statistics*, **4**, 199–212.

Stafford, J.E. (1996). A robust adjustment of the profile likelihood. *Annals of Statistics*, **24**, 336–352.

Stafford, J.E. (2000). Using intersection matrices to identify graphical structure. *Statistics and Computing*, in press.

Stafford, J.E. and Andrews, D.F. (1993). A symbolic algorithm for studying adjustments to the profile likelihood. *Biometrika*, **80**, 715–730.

Stafford, J.E., Andrews, D.F., and Wang, Y. (1993). Symbolic computation: A unified approach to studying likelihood. *Statistics and Computing*, **4**, 235–245.

Stafford, J.E. and Bellhouse, D.R. (1995). A computer algebra for sample survey theory. *Survey Methodology*, **23**, 3–10.

Venables, W.N. (1985). The multiple correlation coefficient and Fisher's. *Aust. J. Statist.*, **27**, 172–182.

Welch, B.L. (1965). On comparisons between confidence point procedures in the case of a single parameter. *Journal of the Royal Statistical Society*, **B27**, 1–8.

Welch, B.L. and Peers, H.W. (1963). On formulae for confidence points based on integrals of weighted likelihoods. *Journal of the Royal Statistical Society*, **B25**, 318–329.

Wishart, J. (1952). Moment coefficients of the k-statistics in samples from a finite population. *Biometrika*, **39**, 1–13.

Withers, C.S. (1983). Asymptotic expansions for distributions and quantiles with power series cumulants. *Journal of the Royal Statistical Society*, **B46**, 389–396.

Wolfram, S. (1996). *The Mathematica Book*, 3rd Edition. Cambridge University Press.

Yong Wang (1992). *Symbolic Computation for Statistics Using Tensors*. Ph.D. thesis, University of Toronto.

Young, A.E. and Daniels, H.E. (1990). Bootstrap bias. *Biometrika*, **77**, 179–185.

Name index

Subject index